The Molecular Biology of Chronic Heart Failure

Colloquium
Digital Library of Life Sciences

This e-book is a copyrighted work in the Colloquium Digital Library—an innovative collection of time saving references and tools for researchers and students who want to quickly get up to speed in a new area or fundamental biomedical/life sciences topic. Each PDF e-book in the collection is an in-depth overview of a fast-moving or fundamental area of research, authored by a prominent contributor to the field. We call these e-books *Lectures* because they are intended for a broad, diverse audience of life scientists, in the spirit of a plenary lecture delivered by a keynote speaker or visiting professor. Individual e-books are published as contributions to a particular thematic **series**, each covering a different subject area and managed by its own prestigious editor, who oversees topic and author selection as well as scientific review. Readers are invited to see highlights of fields other than their own, keep up with advances in various disciplines, and refresh their understanding of core concepts in cell & molecular biology.

For the full list of published and forthcoming Lectures, please visit the Colloquium homepage: www.morganclaypool.com/page/lifesci

Access to Colloquium Digital Library is available by institutional license. Please e-mail info@morganclaypool.com for more information.

Morgan & Claypool Life Sciences is a signatory to the STM Permission Guidelines. All figures used with permission.

Colloquium Series on Genomic and Molecular Medicine

Editor in Chief
Dhavendra Kumar
Institute of Molecular and Experimental Medicine
Cardiff University School of Medicine
University Hospital of Wales
Cardiff, UK

From 1970 onwards, there has been a continuous and growing recognition of the molecular basis of medical practice. Alongside the developments and progress in molecular medicine, new and rapid discoveries in genetics have led to an entirely new approach to the practice of clinical medicine. Until recently the field of genetic medicine has largely been restricted to the diagnosis of disease, offering explanation and assistance to patients and clinicians in dealing with a number of relatively uncommon inherited disorders. However, since the completion of the human genome in 2003 and several other genomes, there is now a plethora of information available that has attracted the attention of molecular biologists and allied researchers. A new biological science of Genomics is now with us, with far reaching dimensions and applications.

During the last decade, rapid progress has been made in new genome-level diagnostic and prognostic laboratory methods, and revealing findings in genomics have led to changes in our understanding of fundamental concepts in cell and molecular biology. It may well be that evolutionary and morbid changes at the genome level could be the basis of normal human variation and disease. Applications of individual genomic information in clinical medicine have led to the prospect of robust evidence-based personalized medicine, and genomics has led to the discovery and development of a number of new drugs with far reaching implications in pharmaco-therapeutics. The existence of Genomic Medicine around us is inseparable from molecular medicine, and it contains tremendous implications for the future of clinical medicine.

The Molecular Biology of Chronic Heart Failure
Dhavendra Kumar
www.morganclaypool.com

ISBN: 9781615045563 paperback

ISBN: 9781615045570 ebook

DOI: 10.4199/C00071ED1V01Y201212GMM003

A Publication in the

COLLOQUIUM SERIES ON GENOMIC AND MOLECULAR MEDICINE

Lecture #3

Series Editor: Dhavendra Kumar, Institute of Molecular and Experimental Medicine, Cardiff University School of Medicine, UK

Series ISSN

ISSN 2167-7840   print
ISSN 2167-7859   electronic

The Molecular Biology of Chronic Heart Failure

Dhavendra Kumar
Institute of Molecular and Experimental Medicine
Cardiff University School of Medicine
University Hospital of Wales
Cardiff, UK

COLLOQUIUM SERIES ON GENOMIC AND MOLECULAR MEDICINE #3

ABSTRACT

The clinical syndrome of chronic heart failure (CHF) is the hallmark of progressive cardiac decompensation, one of the most common chronic medical conditions that affect around 2% of the adult population worldwide irrespective of ethnic and geographic origin (Anonymous). Apart from ischemic heart disease, hypertension, infection, and inflammation, several other etiologic factors account for irreparable and irreversible myocardial damage leading to heart failure (HF). Genetic and genomic factors are now increasingly identified as one of the leading underlying factors (Arab and Liu 2005). These factors may be related to pathogenic alterations (mutation or polymorphism) within specific cardiac genes, mutations in genes incorporating single or multiple molecular pathways (protein families) relevant to cardiac structure and/or function, genetic or genomic polymorphisms of uncertain significance (gene variants, single-nucleotide polymorphisms (SNPs), and copy number variations (CNVs)), and epigenetic or epigenomic changes that influence cardiac gene functions scattered across the human genome. Recent genetic and genomic studies in both systolic and diastolic ventricular dysfunction, the hallmark of CHF, have revealed a number of mutations in genes belonging to specific cardiac protein families. For example, around 200 mutations are now known to exist in around 15 genes coding for several different types of sarcomere proteins (Liew and Dzau 2004). The sarcomere protein family, alone, accounts for the bulk of inherited cardiomyopathies including hypertrophic cardiomyopathy (HCM), dilated cardiomyopathy (DCM), restrictive cardiomyopathy (RCM), and left ventricular (LV) non-compaction (LVNC). In addition, there are several other potentially relevant factors involving different genes and genome-level elements. This article presents a systematic account on the available factual information and interpretations based on genetic and genomic studies in CHF (Liew and Dzau 2004). Genomic and molecular approaches have opened the way for a renewed debate for taxonomy of CHF (Ashrafian and Watkins 2007). The review draws attention to the potential diagnostic and therapeutic implications of genomic and transcriptional profiling in HF and translational genomics research that is likely to permit greater personalization of prevention and treatment strategies to address the complexities of managing clinical HF (Creemers, Wilde et al. 2011).

KEYWORDS

heart failure; genetics, genomics, molecular genetics, molecular pathology; molecular biology; mendelian disease; inherited cardiac disease; cardiomyopathy; atrial fibrillation; personalized medicine; gene mutation; gene (DNA) polymorphisms

Contents

CHAPTER 1

Introduction

Even before the idea for the Human Genome Project was conceived and debated, the concept of association of complex and heterogeneous phenotypes with many genes and related genetic variants had established its place in basic and applied biomedical science. Extensive data and information are now available on the public domain derived from a number of genetic studies on several common and uncommon phenotypes including chronic heart failure (CHF) (PubMed/OMIM). Following the completion of the Human Genome Project, rapid and continued advances in genomic profiling and basic molecular insights have led to a more complete and in-depth revelation of interactions between genes, genomic elements, environment, and affected tissues and organs (Morita, Seidman et al. 2005; Margulies, Bednarik et al. 2009). This has led to a comprehensive understanding of biological mechanisms underlying pathological changes within organs and tissues affected by complex and heterogeneous conditions like heart failure (HF). Two major themes have emerged from these studies—the patients' genetic endowment and the transcriptional changes within the target organ and tissue. Recent studies (MacRae 2010; Velagaleti RS 2010) provide new insights into the mechanisms contributing to the structural changes of the heart and the development of HF. New technological advances have enabled to characterize the patients' genotype and markers of gene expression in detail propelling to new diagnostic and therapeutic opportunities for HF patients (Braunwald 2008; Chaanine, Kalman et al. 2010). In this context, this review focuses on major genetic and genomic aspects of HF highlighting contemporary efforts to achieve greater personalization of prevention and treatment strategies to meet with the challenges of a growing HF epidemic.

CHAPTER 2

Clinical Characteristics of Heart Failure

Most cardiology texts include details on clinical features of acute and chronic heart failure. In this review, only relevant aspects of clinical characteristics are considered that are essential in making a distinction between inherited, acquired, or a combination of both (Gomes, Falcão-Pires et al.). Essentially, HF results from occlusive or non-occlusive pathologies resulting in myocardial dysfunction due to volume or pressure overload. However, in clinical practice, a distinction between occlusive and non-occlusive HF may not always be possible. The clinical syndrome of HF results from different causes and interaction of multiple compensatory and adaptive pathways (Figure 1; Ashrafian and Watkins 2007). Early onset, moderate to severe HF results from a variety of congenital heart disease involving the right and left outflow tracts and inter-ventricular left to right shunt. The most common cause of right ventricular heart failure (RVHF) is LV dysfunction complicating elevated pressure in the pulmonary circulation. The common causes of RVHF include congenital heart disease (for example, Tetralogy of Fallot), primary pulmonary hypertension, and arrythmogenic right ventricular cardiomyopathy (ARVC)/dysplasia and right ventricular infarction. Congenital or acquired valvular stenosis (tricuspid, pulmonary, mitral, and aortic) and systemic and pulmonary hypertension are the main causes of pressure overload manifesting with HF. Cardiac causes of volume overload include left to right shunt in congenital heart anomalies (ventricular and atrioventricular septal defects) and mitral and aortic regurgitation. In addition, systemic causes of volume overload (severe anemia, hypoalubminemia, chronic hepatic and renal decompensation) also contribute to RVHF or congestive HF.

Ischemic heart disease (IHD) is, by far, the most common cause for acute, subacute, or CHF through a clinical course manifested with acute coronary syndrome, acute myocardial infarction, and recurrent ischemic episodes leading to progressive pump failure. The population-specific incidence and prevalence data on IHD is widely available with a consistent pattern in most countries with some correlation to the socioeconomic status. IHD is universally acknowledged to be a disease of the developed world influenced by a rising incidence of obesity and diabetes mellitus. The high prevalence of HF in African-American and people of Afro-Caribbean origin is linked to hypertension. This is further complicated by dietary and lifestyle factors. However, genetic and

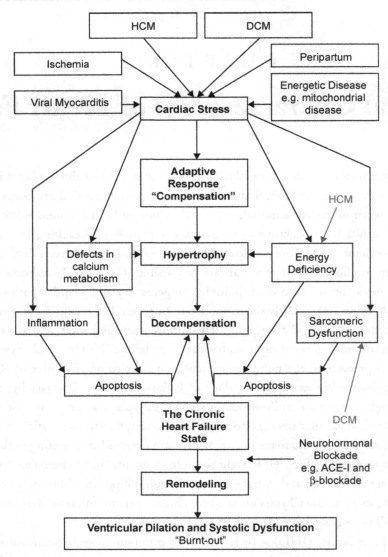

FIGURE 1: Diverse etiologic factors, compensatory and adaptive mechanisms involved in the syndrome of chronic heart failure (modified from Ashrafian and Watkins 2007).

genomic factors are significant and have recently been elucidated by several genomic studies (Seo and Goldschmidt-Clermont 2008).

The infectious causes of HF (for example post-streptococcal rheumatic fever and Chagas' disease) are declining in developed nations but remain endemic in tropical developing and under-developed countries. Viruses (for example, Coxsackie's B3, parvovirus, etc.) are the major suspected culprits for idiopathic dilated cardiomyopathy (IDCM) in which the post-viral inflammation and apoptosis trigger ventricular remodeling and dilation (Liu and Mason 2001). IDCM accounts for 30% of the cases of DCM and is probably molecularly associated with several low- to medium-risk genes and polymorphisms spread across the genome (Osterziel and Perrot 2005).

Non-infectious causes of HF include iatrogenic valvular disease (oral contraceptive pills, toxins). Alcohol abuse may also lead to HF, often presenting acutely following excessive consumption over a relatively short period. HF on presentation in the peri-partum or post-partum period has a variable clinical course from severe pump failure to complete recovery. Stress cardiomyopathy is a rare reversible form of LV dysfunction associated clinically with emotional stress, angiographically with "apical ballooning" and pathophysiologically with excess sympathetic activation (Wittstein, Thiemann et al. 2005). The condition may mimic ST segment elevation myocardial infarction (STEMI) but carries a better prognosis on the removal of stress and emotional factors. Other medical conditions like severe chronic anemia, thyrotoxicosis, and Paget's disease may complicate with high output (volume overload) HF. More recently, the clinical syndrome of HF with preserved ejection fraction (HFPEF) has been described with LV ejection fraction >50% (Bhatia, Tu et al. 2006). A patient with HFPEF is often an elderly female with a history of recurrent atrial fibrillation and hypertension. A precise etiologic relationship with either acquired or genetic factors is unclear for HFPEF.

.   .   .   .

CHAPTER 3

Patterns of Heart Failure

Historically, HF has been described and classified in a number of ways reflecting the knowledge and understanding of the pathophysiology around that time. However, two forms of HF are widely recognized—systolic heart failure (SHF) and diastolic heart failure (DHF). Both forms could result from either volume or pressure overload. SHF is associated with a decreased cardiac output and ventricular contractility, also called systolic dysfunction attributed to a loss of ventricular muscle cells. Systemic hypertension and aortic stenosis remain the major causes for SHF. In classic terms, SHF is progressive, characterized by impaired systolic function and enlargement of one of more ventricles indistinguishable from DCM. In contrast, HCM results in DHF, associated with a combination of increased cardiac muscle stiffness and decreased ventricular contractility. The outcome in both SHF and DHF is dependent upon the pace and degree of myocardial remodeling, a natural phenomenon recognized in any form of myocardial cell damage due to either genetic or non-genetic factors or a combination of both. Myocardial remodeling may involve one or more cardiac-specific protein families discussed later in another section (see Section 2—Pathophysiology of HF).

The New York Heart Association (NYHA) functional classification scheme for HF is a widely used screening tool and assesses the severity of functional limitations of individuals afflicted with HF. The four classes of the NYHA classification are linked to increasing the severity of signs and symptoms and correlate well with the prognosis. However, this classification scheme has certain limitations as diverse pathophysiological processes leading to symptomatic HF may be overlooked (Dunselman, Kuntze et al. 1988). This is addressed by the American College of Cardiology and American Heart Association (ACC/AHA) classification that take into account multiple stages of HF and predisposing factors. Regular updates and guidelines on HF are produced by ACC/AHA. Some of these updates are authoritative sources on the evaluation, management, performance measures, and outcomes on HF (Bonow, Bennett et al. 2005). These guidelines subdivide HF into four stages incorporating preclinical stages, pathophysiologic stages, and clinical recognition of HF (Table 1).

CHF can be either predominantly right sided or left sided. However, a clear distinction may not be possible toward the severe end of the spectrum due to the disruption in peripheral and pulmonary hemodynamics. A number of primary and acquired pulmonary vascular conditions

TABLE 1: ACC/AHA modification of NYHA classification for heart failure

Stage A patients are at high risk for developing heart failure, but have had neither symptoms nor evidence of structural cardiac abnormalities. Major risk factors include hypertension, diabetes mellitus, coronary artery disease and family history of cardiomyopathy. In selected patients, the administration of angio-tensin converting enzyme (ACE) inhibitor is recommended to prevent adverse ventricular remodeling.

Stage B patients have structural abnormalities from previous myocardial infarction. LV dysfunction or valvular heart disease but have remained asymptomatic. Both ACE inhibitors, and beta-blockers are recommended.

Stage C patients have evidence for structural abnormalities along with current or previous symptoms of dyspnea, fatigue and impaired exercise tolerance. In addition to ACE inhibitors and beta-blockers, optimal medical regimen may include diuretics, digoxin, and aldosterone antagonists.

Stage D patients have end-stage symptoms of heart failure that are refractory to standard maximal medical therapy. Such patients are candidates for left ventricular assist devices (LVADs) and other sophisticated maneuvers for myocardial salvage or end-of-life care.

result in pulmonary hypertension leading to right-sided HF. Inherited cardiovascular conditions that characteristically cause pulmonary hypertension include hereditary primary arterial hypertension (HPAH) and hereditary hemorrhagic telangiectasia (HHT), both autosomal dominant disorders with high penetrance and marked clinical variation. The penetrance is close to 100%; however, clinical diagnosis may be missed due to very subtle clinical features in some mutation carriers (heterozygotes). In addition, genetic factors may be important in sporadic PAH and persistent fetal circulation. It is also likely that genetic predisposition may be likely in some acquired form of PAH, for example, recurrent pulmonary thromboembolic episodes, congenital heart disease complicating raised pulmonary arterial and venous hypertension (Eiser–Manger syndrome). Diagnosis and medical management of pulmonary arterial hypertension has remarkably changed with improved clinical outcomes and increased long-term survival. Mutations in genes encoding several proteins in the *bone morphogenic protein receptor–transforming growth receptor–SMAD pathway* are causally linked to HPAH and some other heritable disorders complicating PAH. Diagnostic genetic testing in *BMPR2*, *ALK1*, *endoglin* genes allows early diagnosis and precise identification of the *at-risk* family members. Multi-system life-long clinical surveillance is often necessary in selected family members, either by family history (first-degree relatives) or confirmed by predictive genetic testing. Details on HPAH, HHT, and related disorders are not included in this review and can be found elsewhere (Aldred, Vijayakrishnan et al. 2006; Machado, Eickelberg et al. 2009).

·   ·   ·   ·

CHAPTER 4

Pathophysiology of Chronic Heart Failure

4.1 REFLEX EFFECTS AND NEUROHUMORAL CHANGES

A number of successive pathophysiological changes follow myocardial cell injury or death leading to progressive deterioration of the LV function. These changes are not confined to cardiovascular system alone but involve other body systems notably the skeletal muscles (reflex effects and atrophic muscle changes) (Drexler, Riede et al. 1992), vascular system (reduced peripheral blood flow and endothelial vascular function) (Ino-Oka, Kutsuwa et al. 2001), and lung function (ventilatory abnormalities) (Chua and Coats 1995). Involvement of the skeletal muscles through reflex effects and neurohumoral changes negatively contributes to the development of the syndromes of clinical HF (Figure 2; Coats 2001). The LV damage, whether ischemic or non-ischemic, triggers adverse

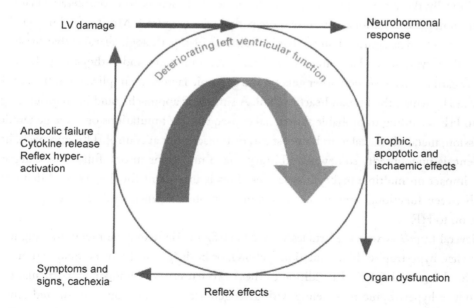

FIGURE 2: Pathophysiological changes in the syndrome of clinical heart failure (adapted with permission from Coats 2001).

neurohormonal responses that exaggerate trophic, apoptotic, and ischemia effects of reduced LV output. This is reflected in progressive deterioration of multiple organ dysfunction. The reflex effects on the skeletal muscle cause fatigue, muscle weakness, and muscle wasting leading to anabolic changes and low-grade inflammation with the release of cytokines and the accumulation of lactate and other toxic metabolites. Inherent changes in any of the molecular cascade pathways would add to the clinical severity due to increased prior genetic predisposition. This would be more evident in monogenic cardiac (inherited cardiomyopathies) and non-cardiac (primary myopathies such as Duchenne and Becker muscular dystrophy) conditions that inadvertently result in CHF.

4.2   MECHANISMS OF CARDIAC CELL DEATH AND CARDIAC REMODELING

The cardiac remodeling is the earliest clinicopathological response to cardiac cell injury or death (Jessup and Brozena 2003). Clinically, this manifests with the syndrome of HF. The cardiac (ventricular) remodeling eventually leads to one of the two distinct morphologies—LV hypertrophy (increased wall thickness without chamber expansion) or dilation (normal or thinned walls with enlarged chamber volumes), associated with specific hemodynamic changes (Figure 3). Systolic function is normal in hypertrophic remodeling; however, diastolic relaxation is impaired. In contrast, diminished systolic function characterizes dilated remodeling. Early clinical recognition of these cardiac findings is important to diagnose either HCM or DCM.

Clinically, these may appear as specific diagnoses, but there is now considerable evidence that many different gene mutations can cause these pathologies (Figure 4; Morita, Seidman et al. 2005). Both types can be distinguished on the basis of distinct histopathologic features that further delineate several subtypes of cardiac remodeling. Apart from mutations and pathogenic polymorphisms in distinct cardiac genes, several other genetic and genomic factors are implicated in the multiplicity of pathways by which the human heart can fail. A simplistic approach could be to group all genetic factors in HF according to probable functional consequences of mutations on force generation and transmission, metabolism, calcium homeostasis, or transcriptional control. This approach, however convenient, is undoubtedly somewhat arbitrary. Gene mutations in one functional category may have an impact on multiple myocyte processes. This is important for the eventual delineation of signals between functional groups that result in, or protect against, cardiac decompensation and progression to HF.

Several http://www.jci.org/articles/view/24351/figure/1human gene mutations (Figure 4) can cause cardiac hypertrophy (blue), dilation (yellow), or both (green) (Morita, Seidman et al. 2005; Alcalai, Seidman et al. 2008). In addition to these two patterns of remodeling, particular gene defects produce hypertrophic remodeling with glycogen accumulation (pink) or dilated remodeling with fibrofatty degeneration of the myocardium (orange). Sarcomere proteins denote β-myosin

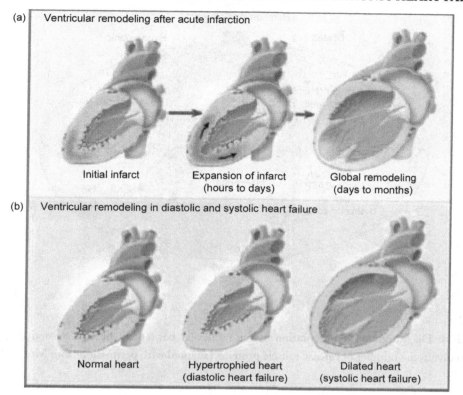

FIGURE 3: Ventricular remodeling after infarction (a) and in diastolic heart failure (b) (adapted from Jessup and Brozena 2003).

heavy chain, cardiac troponin T, cardiac troponin I, α-tropomyosin, cardiac actin, and titin. The metabolic/storage proteins denote AMP-activated protein kinase γ subunit, LAMP2, lysosomal acid α 1,4-glucosidase, and lysosomal hydrolase α-galactosidase A. The Z-disc proteins denote MLP and telethonin. Dystrophin-complex proteins denote δ-sarcoglycan, β-sarcoglycan, and dystrophin. Ca^{2+} cycling proteins denote PLN and RyR2. Desmosome proteins denote plakoglobin, desmoplakin, and plakophilin-2.

Functional analysis of human gene mutations that cause HCM and DCM provides information about the triggers of cardiac remodeling. In addition to understanding the molecular-signaling cascades, gene-expression profiling may offer the opportunity to define the precise pathways implicated in HF. This can be achieved by a comprehensive data set of the transcriptional and proteomic profiles associated with precise gene mutations. Despite the plethora of information available, bioinformatic assembly of data and deduction of pathways should be feasible and productive for defining shared or distinct responses to signals that cause cardiac remodeling and HF. Ultimately, this data

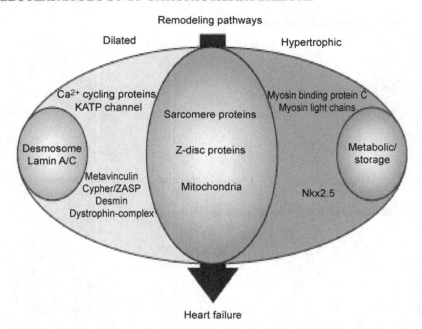

FIGURE 4: Diagrammatic representation of the molecular basis of myocardial remodeling in heart failure involving the number of cardiac muscle proteins (adapted with permission from Morita, Seidman et al. 2005).

set in humans is a desirable goal; however, information on clinical variables and laboratory analysis may pose considerable difficulties that can be more readily addressed by a study of animal models with HF. It is anticipated that these resources will eventually help in developing HF management strategies for minimizing ventricular hypertrophy, reduce myocyte death, and diminish myocardial fibrosis (Morita, Seidman et al. 2005).

4.3   FETAL GENE REACTIVATION IN MYOCARDIAL REMODELING

It is widely believed that cardiac remodeling is the natural consequence of a cardiac cell injury resulting in CHF. Several molecular changes follow cardiac cell injury including reactivation of the fetal gene program triggering pathological changes in the myocardium associated with progressive myocardial dysfunction (Hunter and Chien 1999; Hoshijima and Chien 2002). Recent studies provide evidence for fetal cardiac microRNAs (miRNAs) as the key regulators of gene expression contributing to the transcriptional changes observed in HF (Thum, Gross et al. 2008).

miRNAs are recently discovered regulatory molecules consisting of ~22 non-coding nucleotides that regulate expression by hybridization to messenger RNAs (mRNAs) with the consequence of mRNA degradation or translational inhibition of miRNAs (Ambros 2001)). More than 350 human miRNAs have been identified (http://microrna.sanger.ac.uk), each targeting numerous different RNA transcripts with the consequence of gene silencing. Databases and computation approaches have been developed to predict target genes of miRNAs; however, only few have so far been validated (Wang and Wang 2006).

A number of biological processes are regulated by miRNAs (Esquela-Kerscher and Slack 2006) including cancer and developmental processes (Wienholds, Kloosterman et al. 2005; Giralder and Hammond 2005). Recent evidence suggests that miRNAs are also implicated in controlling cardiac development and hypertrophy (van Rooij, Sutherland et al. 2006; van Rooij, Sutherland et al. 2007; Sayed, Rane et al. 2008). This is supported by the fact that in the fruit fly, miRNA (*miR-1*) targets transcripts encoding the Notch ligand delta, thus, regulating the expansion of cardiac and muscle progenitor cells (Kwon, Han et al. 2005). Accumulation of *miR-1* in the heart impairs the pool of proliferating ventricular myocytes (Zhao, Samal et al. 2005) and results in the dysregulation of cardiogenesis, whereas targeted deletion of *miR-1* results only in the dysregulation of cardiogenesis (Zhao and Srivastava 2007).

It is believed that alterations of specific miRNAs probably contribute to the reactivation of fetal gene programs in human HF (Thum, Gross et al. 2008). Using microarray and miRNA stem loop real-time polymerase chain reaction (RT-PCR) technology, the group investigated miRNA profiles as well as the cardiac mRNA transcriptome in LV tissue from patients with end-stage HF [in comparison with tissue from healthy adult and fetal human hearts. There are potential target genes of regulated miRNAs validated by database searches and simultaneous transcriptome and quantitative RT-PCR analyses. Profound alterations of miRNA expression were observed in failing hearts. These changes closely mimicked the miRNA expression pattern conserved in fetal cardiac tissue (Tatsuguchi, Seok et al. 2007). Bioinformatic analysis demonstrated a striking concordance between regulated messenger RNA expression in HF and the presence of miRNA binding sites in the respective 3 untranslated regions. Messenger RNAs upregulated in the failing heart contained preferentially binding sites for downregulated microRNAs and vice versa. Reexpression of miRNAs in cultured neonatal or adult cardiomyocytes result in the activation of gene programs and, consequently, morphological changes as observed during HF. Future identification of miRNA actions and functions will substantially improve our understanding of cardiovascular biology. Development of drugs and molecules that specifically regulate cardiac miRNAs with subsequent normalization of altered target expression may lead to novel treatments for HF (Czech 2006). Thus, targeting miRNAs may finally open novel mechanistic treatment concepts for HF prevention and therapy.

· · · ·

CHAPTER 5

Molecular Pathology in Chronic Heart Failure

5.1   MAJOR CARDIAC MUSCLE GENES

Conventionally, most clinical geneticists, genetic counselors and clinical cardiologists focus on mutations in major cardiac genes implicated in cardiomyopathies. Despite limitations, the pathogenicity of such mutations is usually judged on the basis of phenotypic association. Databases of such mutations exist (The Seidman Lab, http://genepath.med.harvard.edu/~seidman/ and Human Mutation Database, www.hmdb.cf.ac.uk) and are often referred to when preparing a molecular genetic diagnostic report. Increasingly, most diagnostic laboratories regularly report missense mutations, polymorphic variants within an exon, splice site variants, and intronic polymorphisms. While it is tempting to correlate these with the phenotype, computational (Insilico) analysis and gene expression (mRNA studies) analysis fail to provide an evidence for pathogenicity.

It is now acknowledged that the number of genes spread across the human genome is limited to around 25,000 compared to 100,000 to 150,000 protein transcripts based on the analysis of expressed sequence tags (ESTs) (Fehlbaum, Guihal et al. 2005). Most protein transcripts happen to be isoforms of a major protein molecule. This discrepancy between the total number of genes and protein isoforms is explained by the phenomenon of alternative RNA splicing, post-translational biochemical modifications, and some other biochemical modifications. The basic dynamics of alternate splicing is illustrated in Figure 5. An alternatively spliced transcript may encode a protein that is missing a specific functional domain, thereby, altering its activity. The frequency of specific splice variants can be defined by disease states, routes of cellular differentiation and phenotype, shifts in metabolic pathways, occurrence of SNPs in splice/acceptor motifs, and other physiological stimuli (Margulies, Bednarik et al. 2009).

In terms of cardiovascular genomics, splicing variants are reported in several transcripts, including troponin T (Maass, Ikeda et al. 2004) and phosphodiesterases (Omori and Kotera 2007). It is likely that alternative splicing in troponin T gene may be causally related to sudden cardiac death (SCD) in HCM (Maass, Ikeda et al. 2004). This mechanism probably accounts for variable

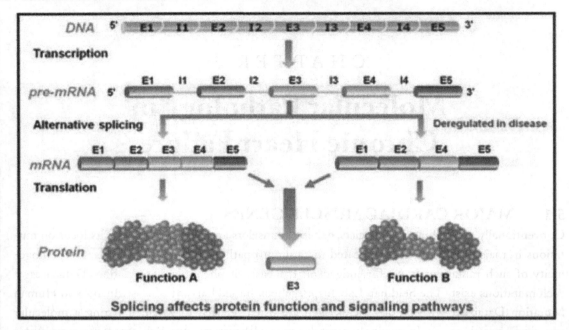

FIGURE 5: Alternative mRNA splicing between transcription and translation processes. Note that the alternative splicing at exon 3 level results in two protein isoforms with differing functional outcomes triggered down the signaling pathways (adapted from Margulies, Bednarik et al. 2009).

phenotypic outcomes in the phosphodiesterase (PDE) gene family where alternative splicing results in multiple mRNA transcripts producing proteins that may differ substantially from one another, even to the extent of having opposing effects (Fehlbaum, Guihal et al. 2005; Omori and Kotera 2007). This is shown in experimental settings using the alternative splicing in failing myocardium (Rose, Armoundas et al. 2005). An impressive account of a pre-proliferative and anti-apoptotic splice variant of the cell cycle kinase transcript associated with HF is provided that is downregulated subsequent to myocardial infarction (Qiu, Dai et al. 2008). In this case, two splice variants differ from one another in terms of protein-protein interactions, substrate specificity, and regulation of the cell cycle. In the future, the ability to perform genome-wide splice variant profiling may extend our understanding of the functional importance of alternative splicing in the development and regulation of HF (Bracco, Throo et al. 2006).

One of the fundamental functions of cardiac myocytes is the generation of contractile force by the sarcomere and its transmission to the extracellular matrix. Inadequate performance in either component initiates cardiac remodeling (hypertrophy and dilation), produces symptoms, and leads to HF. Molecular genetic studies in inherited cardiomyopathies have led to the identification and cloning of several genes that encode a number of cardiac muscle proteins (Table 2; Liew 2004;

SYMBOL	CHROMOSOME	GENE PRODUCT	CARDIOMYOPATHY TYPE
ACTC	15q11–14	Cardiac muscle α-actin	Hypertrophic and dilated
ABCC9	12p12.1	Member 9 of the superfamily C of ATP- binding cassette (ABC) transporters	Dilated
CSRP3, MLP	11p15.1	Cysteine- and glycine-rich protein 3	Dilated
DES	2q35	Desmin	Dilated
DSP	6p24	Desmopfakin	Dilated
LMNA	1q21.2–21.3	Lamin A/C	Dilated
VCL	10q22.1–q23	MetavInculin	Dilated
MYBPC3	11p11.2	Cardiac myosin-binding protein C	Hypertrophic and dilated
MYH6	14q12	Cardiac muscle a-isoform of myosin heavy chain (heavy polypeptide 6)	Hypertrophic
MYH7	14q12	Cardiac muscle a-isoform of myosin heavy chain (heavy polypeptide 7)	Hypertrophic and dilated
MYL2	12q23–24.3	Myosin regulatory light chain associated with cardiac myosin-β (or slow) heavy chain	Hypertrophic

TABLE 2: The genetic basis of cardiomyopathies (Liew and Dzau 2004)

TABLE 2: (*continued*)

SYMBOL	CHROMOSOME	GENE PRODUCT	CARDIOMYOPATHY TYPE
MYL3	3p21.2–21.3	Myosin light chain 3	Hypertrophic
PLN	6q22.1	Phospholamban	Dilated
PRKAG2	7q35–36	γ2 non-catalytic subunit of AMP-activated protein kinase	Hypertrophic
SGCB	4q12	β-Sarcoglycan (43kDa dystrophin-associated glycoprotein)	Dilated
SGCD	5q33–34	δ-Sarcoglycan (35kDa dystrophin-associated glycoprotein)	Dilated
TAZ, G4.5	Xq28	Tafazzin	Dilated
TTN	2q31	Titin	Hypertrophic and dilated
TCAP	17q12	Titin-cap	Dilated
TPM1	15q22.1	Tropomyosin 1 (α)	Hypertrophic
TNNI3	19q13.4	Troponin I, a subunit of the troponin complex of the thin filaments of striated muscle	Hypertrophic
TNNT2	1q32	Cardiac isoform of troponin T2, tropomyosin-binding subunit of the troponin complex	Hypertrophic and dilated

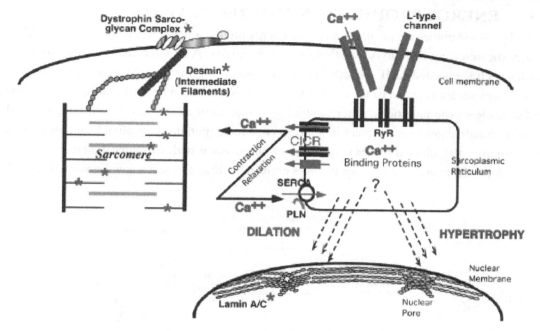

FIGURE 6: Major genes and proteins in cardiac myocytes involved in remodeling (adapted from Seidman and Seidman 2001).

Creemers, Wilde et al. 2011; Esther, Hoberman et al. 2011). Some of these genes are specific for a particular type of inherited cardiomyopathy; however, most overlap with different phenotypes. Despite similarities in function, these may be grouped in different functional categories including force generation and propagation, energy production and regulation, calcium cycling and transcriptional regulators (Figure 6; Table 2; Seidman and Seidman 2001; Morita, Seidman et al. 2005).

5.2   FORCE GENERATION AND PROPAGATION

The functional unit involved in force generation and propagation is made up of sarcomere proteins (thick and thin filaments), dystrophin-associated glycoproteins (intermediate filaments), intercalated and Z-disc proteins, lamin A/C nuclear membrane proteins, and desmosome junctional cell membrane proteins (Figure 4). Human mutations in the genes encoding these proteins trigger cardiac remodeling (HCM or DCM). While the clinical outcome is HF, the histopathological changes, hemodynamic and biophysical profiles are different in HCM and DCM suggesting that distinct molecular processes are probably involved. Detailed account of genes and mutations disrupting the structure and function of these proteins is reviewed elsewhere (Morita, Seidman et al. 2005).

5.3   ENERGY PRODUCTION AND REGULATION

Several protein families are involved in cardiac energy production and regulation (Figure 7). These include the mitochondrial DNA (mtDNA)-encoded proteins (ATP and oxidative phosphorylation) and the nuclear-encoded AMP-activated protein kinases. While many of the protein components of these complexes are encoded by the nuclear genome, 13 are encoded by the mitochondrial genome. Unlike nuclear gene mutations, mitochondrial gene mutations exhibit matrilineal inheritance. As the mitochondrial genome is present in multiple copies and mutations are often heteroplasmic, the distribution of the affected copies is random and inconsistent with the clinical picture. However, energy-dependent tissues and organs are more prone to early and/or severe involvement, such as the

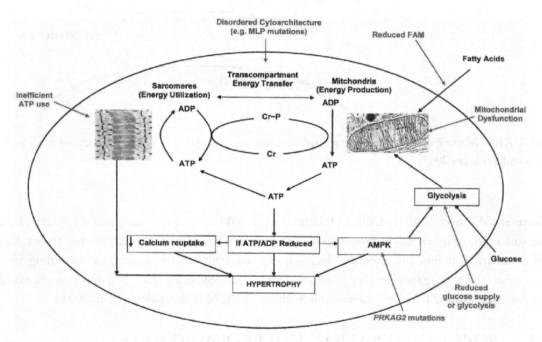

FIGURE 7: Cardiac energy production and regulation; the phenotype of hypertrophic cardiomyopathy (HCM) can arise from: (1) excessive energy wastage by inefficient ATP use (for example, by aberrant sarcomeres); (2) inadequate energy production (for example, from poorly functioning mitochondria); (3) inadequate metabolic substrates (reduced glucose uptake or glycolysis); (4) failure to transfer energy across cellular compartments owing to cytoarchitectural defects as exemplified by muscle LIM protein (MLP) mutations; or (5) aberrant signaling of energy deficiency (e.g., with AMP-activated protein kinase (AMPK) mutations). The final common path for these diverse defects is energy deficiency and ensuing hypertrophy (ADP—adenosine diphosphate; AMP—adenosine monophosphate; ATP—adenosine triphospate; Cr—creatine; FAM—fatty acid metabolism).

heart, nervous system, and sensory organs (e.g., eyes and inner ear). HCM and/or DCM feature in some selected mitochondrial syndromes such as the Kerans–Sayre syndrome, ocular myopathy, mitochondrial encephalopathy with lactic acidosis and stroke-like episodes (MELAS), and myoclonic epilepsy with ragged-red fibers (MERFF) (Chinnery). There is some evidence that particular mitochondrial DNA (mtDNA) mutations can produce predominant or exclusive cardiac disease, in particular, DCM (Limongelli, Tome-Esteban et al. 2010).

In addition to mitochondrial genome-mediated complexes, several nuclear gene-encoded proteins play a key role in cardiac cellular metabolism and, thus, in HF. Mutations in genes encoding protein kinases, for example, the γ2 subunit of the AMP-activated protein kinase (*PRKAG2*), α-galactosidase A (*GLA*), and lysosome-associated membrane protein-2 (*LAMP2*), can cause profound myocardial hypertrophy in association with electrophysiologic defects (Liew and Dzau 2004; Schramm, Fine et al. 2012). Cardiac histopathology reveals that, unlike sarcomere gene mutations, which cause hypertrophic remodeling, the mutations in *PRKAG2*, *LAMP2*, and *GLA* accumulate glycogen in complexes with protein and/or lipids, thereby defining these pathologies as storage cardiomyopathies. Progression from hypertrophy to HF is particularly common and occurs earlier with *LAMP2* mutations than with other gene mutations that cause metabolic cardiomyopathies. As both *GLA* and *LAMP2* are encoded on chromosome X, disease expression is more severe in men, but heterozygous mutations in women are not entirely benign, perhaps due to X inactivation that equally extinguishes a normal or mutant allele. The cellular and molecular pathways that produce either profound hypertrophy or progression to HF from *PRKAG2*, *GLA*, or *LAMP2* mutations are incompletely understood. While accumulated by-products are likely to produce toxicity, animal models indicate that mutant proteins cause far more profound consequences by changing cardiac metabolism and altering cell signaling. This is particularly evident in *PRKAG2* mutations that increase glucose uptake by stimulating the translocation of the glucose transporter GLUT-4 to the plasma membrane, increase hexokinase activity, and alter the expression of signaling cascades (Sternick, Oliva et al. 2011). The cooccurrence of electrophysiologic defects in metabolic mutations raises the possibility that pathologic cardiac conduction and arrhythmias contribute to cardiac remodeling and HF in these gene mutations.

5.4   CALCIUM CYCLING

Abnormalities in myocyte calcium homeostasis are shown to be important in molecular and cellular mechanisms in HF and SCD (Blayney and Lai 2009). Protein and RNA levels of key calcium modulators are altered in acquired and inherited forms of HF, and human mutations in molecules directly involved in calcium cycling have been found in several cardiomyopathies (Figure 8). Calcium enters the myocyte through voltage-gated L-type Ca^{2+} channels triggering the release of calcium from the sarcoplasmic reticulum (SR) via the type-2 ryonidine receptor (RyR2); RyR2 function is stabilized

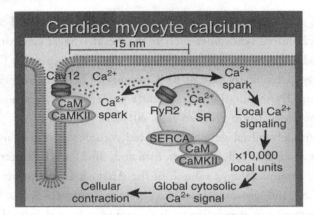

FIGURE 8: Cardiac myocyte calcium. Human mutations affecting Ca^{2+} cycling proteins. Sarcomere contraction is initiated through Ca^{2+} entering through L-type channels (LTCC) triggering Ca^{2+} release (CICR) from the SR via the RyR2; relaxation occurs with SR Ca^{2+} reuptake through the SERCA2a. Calstabin2 coordinates excitation and contraction by modulating RyR2 release of Ca^{2+}. PLN, an SR transmembrane inhibitor of SERCA2a modulates Ca^{2+} reuptake. Dynamic regulation of these molecules is effected by PKA-mediated phosphorylation. Ca^{2+} may further function as a universal signaling molecule, stimulating Ca^{2+}-calmodulin and other molecular cascades (modified from Blayney and Lai 2009).

by the FK506-binding protein (FKBP12.6; calstabin2) preventing aberrant calcium release during the relaxation phase of the cardiac cycle (Figure 8). The stimuli that phosphorylate RyR2 (such as exercise) by protein kinase A (PKA) dissociate calstabin2 from the receptor, thereby increasing calcium release and enhancing contractility. At low concentrations of intracellular calcium, troponin I and actin interactions block actomyosin ATPase activity; increasing levels foster calcium binding to troponin C, which releases troponin I inhibition and stimulates contraction. Cardiac relaxation occurs when calcium dissociates from troponin C, and intracellular concentrations decline as calcium reuptake into the SR occurs through the cardiac sarcoplasmic reticulum Ca^{2+}-ATPase pump (SERCA2a). Calcium reuptake into SR is regulated by phospholamban (PLN), an inhibitor of SERCA2a activity that, when phosphorylated, dissociates from SERCA2a and accelerates ventricular relaxation (Ramay, Liu et al. 2011).

Mutations in RyR2 are associated with arrhythmogenic ARVC and catecholaminergic polymorphic ventricular tachycardia (CPVT). Both are monogenic cardiovascular disorders and feature HF as a major clinical outcome. In this context, mutations in *calsequestrin 2*, an important regulator of RyR2, also result in CPVT. Human *PLN* mutations cause familial DCM and HF (Morita, Seidman et al. 2005). The functional consequence of the *PLN* mutation was predicted to be a constitutive inhibition of SERCA2a, a result confirmed in transgenic mice expressing mutant, but not wild-type, PLN protein. In mutant transgenic mice, calcium transients were markedly prolonged, myocyte relaxation was delayed, and these abnormalities were unresponsive to β-adrenergic stimulation (Morita, Seidman et al. 2005). Profound biventricular cardiac dilation and HF developed in mutant mice, providing clear evidence of the detrimental effects of protracted SERCA2a inhibition due to excess PLN activity. The collective lesson from human *PLN* mutations appears to be that too little or too much PLN activity is bad for long-term heart function. Acquired causes of HF are also characterized by a relative decrease in SERCA2a function due to excessive PLN inhibition.

Cardiac myocytes undergo considerable changes in metabolism and membrane excitability as a result of hypoxia and ischemia. One of the potassium ion channels, cardiac ATP-sensitive potassium channels (K_{ATP} channels), is shown to be critical in decoding metabolic signals to maximize cellular functions during stress adaptation (Morita, Seidman et al. 2005). K_{ATP} channels are multimeric proteins containing the inwardly rectifying potassium channel pore (Kir6.2) and the regulatory SUR2A subunit, an ATPase-harboring, ATP-binding cassette protein. Recently, human mutations in the regulatory SUR2A subunit (encoded by *ABCC9*) were identified as a cause of DCM and HF (Morita, Seidman et al. 2005). These mutations reduced ATP hydrolytic activities rendering the channels insensitive to ADP-induced conformations and, thus, interfering with the channel opening and closure. As K_{ATP}-null mouse hearts have impaired response to stress and are susceptible to calcium overload (Morita, Seidman et al. 2005), some of the pathophysiology of human K_{ATP} mutations (DCM and arrhythmias) may reflect calcium increases triggered by myocyte stress.

5.5   CARDIAC TRANSCRIPTIONAL REGULATORS

Studies in the molecular regulation of the cardiac gene transcription have led to the identification of many key molecules that orchestrate physiologic expression of proteins involved in force production and transmission, metabolism, and calcium cycling. It is likely that mutations in these gene complexes cause cardiac remodeling. However, so far, transcriptional regulation of some of these proteins has not yet been identified as primary causes of HF. This is probably due to reduced reproductive fitness as transcription factor gene mutations may be lethal. The consequences of transcription factor gene mutations may be so pleiotropic that these cause systemic rather than single-organ disease. While many other explanations may be relevant, the few human defects discovered in transcriptional regulators that cause HF provide an important opportunity to understand molecular mechanisms for HF.

Molecular mechanisms that activate or repress cardiac gene transcription have led to the identification of several key molecules that directly or indirectly lead to cardiac remodeling. While human mutations in these genes have not been identified, these molecules are excellent candidates for triggering cell responses to structural protein gene mutations. Important cardiac transcription regulators include Nkx2.5, NFAT3, and GATA4. The homeodomain-containing transcription factor Nkx2.5, a vertebrate homolog of the *Drosophila* homeobox gene *tinman*, is one of the earliest markers of mesoderm. Multiple human dominant *Nkx2.5* mutations have been identified as causing primarily structural malformations (atrial and ventricular septation defects) accompanied by atrioventricular conduction delay, although cardiac hypertrophic remodeling has also been observed (Schott, Benson et al. 1998). Although the mechanism for ventricular hypertrophy in humans with *Nkx2.5* mutations is not fully understood, the pathology is unlike that found in HCM, which perhaps indicates that cardiac hypertrophy is a compensatory event. Heterozygous *Nkx2.5*$^{+/-}$ mice exhibit only congenital malformations with atrioventricular conduction delay (Tanaka, Berul et al. 2002). Remarkably, however, transgenic mice expressing *Nkx2.5* mutations develop profound cardiac conduction disease and HF and exhibit increased sensitivity to doxorubicin-induced apoptosis (Toko, Zhu et al. 2002), which suggests that this transcription factor plays an important role in postnatal heart function and stress response. Recent attention has also been focused on Hop, an atypical homeodomain-only protein that lacks DNA-binding activity. Hop is expressed in the developing heart, downstream of Nkx2-5. While its functions are not fully elucidated, Hop can repress serum response factor-mediated (SRF-mediated) transcription. Mice with *Hop* gene ablation have complex phenotypes. Approximately half of *Hop*-null embryos succumb during mid-gestation with poorly developed myocardium; some have myocardial rupture and pericardial effusion. Other *Hop*-null embryos survive to adulthood with apparently normal heart structure and function. Cardiac transgenic overexpression of epitope-tagged Hop causes hypertrophy, possibly by recruitment of class I HDACs that may inhibit anti-hypertrophic gene expression (Chen, Kook et al. 2002; McKinsey and Olson 2004).

Activation of calcineurin (Ca^{2+}/calmodulin-dependent serine/threonine phosphatase) results in dephosphorylation and nuclear translocation of nuclear factor of activated T cells 3 (NFAT3), which in association with the zinc finger transcription factor GATA4, induces cardiac fetal gene expression. Transgenic mice that express activated calcineurin or NFAT3 in the heart develop profound hypertrophy and progressive decompensation to HF (Molkentin, Lu et al. 1998), responses that were prevented by pharmacologic inhibition of calcineurin. Although these data implicated NFAT signaling in hypertrophic HF, pharmacologic inhibition of this pathway fails to prevent hypertrophy caused by sarcomere gene mutations in mice and even accelerates disease progression to HF (Fatkin, McConnell et al. 2000). Mice lacking calsarcin-1, which is localized with calcineurin to the Z-disc, showed an increase in Z-disc width, marked activation of the fetal gene program, and exaggerated hypertrophy in response to calcineurin activation or mechanical stress, which suggests that calsarcin-1 plays a critical role in linking mechanical stretch sensor machinery to the calcineurin-dependent hypertrophic pathway (Frey, Barrientos et al. 2004).

In the context of calcineurin, histone deacetylases (HDACs) are emerging as important regulators of cardiac gene transcription. The discovery that HDAC kinase is stimulated by calcineurin (Zhang, McKinsey et al. 2002) implicates crosstalk between these hypertrophic signaling pathways. Class II HDACs (4/5/7/9) bind to the cardiac gene transcription factor MEF2 and inhibit MEF2-

target gene expression. Stress-responsive HDAC kinases continue to be identified but may include an important calcium-responsive cardiac protein, calmodulin kinase. Kinase-induced phosphorylation of class II HDACs causes nuclear exit, thereby releasing MEF2 for association with histone acetyltransferase proteins (p300/CBP) and activation of hypertrophic genes. Mice deficient in HDAC9 are sensitized to hypertrophic signals and exhibit stress-dependent cardiac hypertrophy.

Other relevant transcriptional regulator candidates are PPARα and RXRα. PPARα plays important roles in transcriptional control of metabolic genes, particularly those involved in cardiac fatty acid uptake and oxidation. Mice with cardiac-restricted overexpression of *PPARα* replicate the phenotype of diabetic cardiomyopathy (Finck, Lehman et al. 2002). Heterozygous *PPARγ*-deficient mice, when subjected to pressure overload, developed greater hypertrophic remodeling than wild-type controls, implicating the PPARγ-pathway as a protective mechanism for hypertrophy and HF (Asakawa, Takano et al. 2002). Retinoid X receptor α (RXRα) is a retinoid-dependent transcriptional regulator that binds DNA as an RXR/retinoic acid receptor (RAR) heterodimer. *RXRα*-null mice die during embryogenesis with hypoplasia of the ventricular myocardium. In contrast, overexpression of RXRα in the heart does not rescue myocardial hypoplasia but causes DCM (Colbert 2002).

5.6 GENES AND POLYMORPHISMS IN MAJOR CASCADE PATHWAYS

Apart from genes with a major effect, there are several genes and associated gene polymorphisms that are considered medium- to low-risk alleles. These often function in conjunction with other genes encoding proteins within a defined molecular cascade pathway. These cardiac pathways determine and regulate a number of specific biophysical properties. The phenotypic variation between individuals is attributed to genetic variation ascribed to genetic polymorphisms. Genetic polymorphisms are randomly distributed with a population frequency of around 1% and may take several generations to represent a population group. However, certain polymorphisms are present in unequal proportions and may offer some advantage against selection pressure, for example, SCD in a polygenic trait like CHF. Similarly, this may carry a selection disadvantage.

Several factors are linked to adverse outcomes such as SCD in patients with CHF. High risk for SCD in HF of undetermined nature could be attributed to mutations in low-risk genes and gene polymorphisms (Darbar 2010). Malignant ventricular arrhythmia is the final pathophysiological endpoint in CHF resulting in SCD. Similar mechanisms may also result in SCD without a previously known cardiac phenotype. Examples include development of polymorphic ventricular tachycardia upon exposure to a QT-prolonging drug, the so-called "acquired long QT syndrome."

·　·　·　·　·

CHAPTER 6

Genes Linked to Inherited and Acquired Forms of Heart Failure

Most genetic cardiomyopathies eventually result in HF. Thus, an important question is whether mutations in genes implicated in Mendelian cardiomyopathies also result in acquired HF. So far, 26 genes are known in which mutations have been reported to cause genetic forms of HF (Table 3). These include genes for which expression changes have also been described in acquired forms of HF. This is best exemplified by familial dilated cardiomyopathy (DCM) in which out of the 18 genes that are well-established genetic causes in DCM, 8 or 9 are also dynamically regulated in HF. Strikingly, no mutation has yet been reported in genes that are known to be regulated in acquired HF—such as, natriuretic peptide precursor (*NPPA*), *NPPB*, sarco/endoplasmic reticulum Ca$_2$-ATPase (*SERCA*), *GATA4*, and *GATA6*, which are upregulated, and connective tissue growth factor (*CTGF*), peroxisome proliferator-activated receptor-α (*PPARA*) and *PPARB* genes, which are downregulated. There is, therefore, no evidence to suggest that mutations in the genes that always respond to HF actually cause HF. For example, both loss-of-function and gain-of-function mutations in PLN gene have been suggested to cause HF. It is remarkable that no mutation in *SERCA*, which is inhibited by PLN, has been described (Hovnanian 2008). Missense mutations in *SERCA* cause an autosomal-dominant skin disease (Darier's disease) but these patients do not manifest any heart disease (Dhitavat, Dode et al. 2003; Mayosi, Kardos et al. 2006), showing that *SERCA2* is haplosufficient for cardiac muscle function, as has, in fact, been shown in mice (Shull, Okunade et al. 2003).

GENE NAME AND FUNCTIONAL ROLE	GENE EXPRESSION CHANGES IN HEART FAILURE	ISOFORM SWITCH IN THE FAILING HEART
TABLE 3: Genes linked to both genetic (Mendelian) and acquired forms of heart failure (Creemers, Wilde et al. 2011)		
Structural		
MYH6	↓	MYH6 to MYH7
*MYH7**	↑	MYH6 to MYH7
TNNT1, TNNC1* and TNNI1**	↑	No change
Tropomyosin*	↑	TMP1κ
MYBPC3	No change	No change
*ACTC1**	↑	No change
Sarcoglycan, delta*	No change	No change
Dystrophin*	Unknown	No change
Desmin*	↑	No change
Metavinculin	Unknown	No change
Muscle LIM protein	↓	No change
Actinin, alpha	↑	ACTN1 to ACTN2
Titin*	No change	N2BA to N2B
Lamin A/C*	↓	No change
Energy production		
Succinate dehydrogenase complex*	Unknown	No change

GENE NAME AND FUNCTIONAL ROLE	GENE EXPRESSION CHANGES IN HEART FAILURE	ISOFORM SWITCH IN THE FAILING HEART
TABLE 3: (*continued*)		
Calcium handling and ion channels		
Phospholamban*	↑	No change
SUR2A	Unknown	No change
*SCN5A**	↓	No change
Other		
Ankyrin repeat domain 1	↑	No change
Thymopoietin	Unknown	No change
RNA-binding motif 20*	Unknown	No change
LIM-binding domain 3*	Unknown	No change
Eyes absent 4	Unknown	No change
Tafazzin*	Unknown	No change

ACTC1, actin, alpha, cardiac muscle 1; *MYBPC*, myosin binding protein C; *MYH*, myosin heavy chain; *SCN5A*, sodium channel, voltage-gated, type V, alpha subunit; *SUR2A*, sulphonylurea receptors 2a; *TNNC1*, troponin C type l; *TNNI1*, troponin I type l; *TNNT1*, troponin T type 1. *Genes marked by an asterisk represent genes for which a causal role has been shown in large pedigrees (*n* > 3 affected) and/or in independent reports. If not specifically reported, changes in expression have been obtained using NCBI-GEO data sets. A fully referenced version of this table is provided in Supplementary Information S1 (table).

.　　.　　.　　.

CHAPTER 7

Genomic Perspectives in Chronic Heart Failure

While genetics is the study of "single genes and their effects," genomics refer to "the functions and interactions of all the genes in the genome" (Guttmacher and Collins 2002). This distinction is not simply a quantitative expression but also implies that genetic information at one locus is modified by information at many other loci and their interaction with non-genetic factors (Khoury 2003). Following the completion of the human genome and several other genomes, tremendous technical advances have led to the availability of a number of sophisticated fast high-throughput genomic techniques that now allow efficient individual genomic profiling. Genomic profiling is still in its infancy but is acknowledged to be the corner stone for genome-based personalized and evidence-based medical practice (Kumar 2011).

In simple terms, genomic profiling in an individual with a complex disease can be undertaken at the whole genome level, all exon sequences (exomic profiling) and transcriptional sequences (gene-expression profiling). The whole genome profiling will reveal polymorphisms spread across the genome. These include SNPs and CNVs. A number of commercially available microarray platforms are now available (Illumina, Affymetrix, Roche-Nimblegen, etc.) with specific computational software and equipment that offer SNPs and CNVs profiling in a relatively short time. However, clinical utilization of this approach has not yet materialized, except for some correlation in children with developmental delay and unusual physical features (www.decipher.sanger.ac.uk).

7.1 GENOMIC PROFILING

Genomic profiling is increasingly moving into the clinical arena (Table 4). The focus is largely to identify SNPs and CNVs that may collectively offer clinically useful diagnostic power in making therapeutic decisions, assessing the prognosis, early detection, and prevention. Results of several genome-wide association studies (GWAS) in a number of complex phenotypes have been either disappointing or produced evidence of low clinical utility (Donnelly 2008; McCarthy, Abecasis et al. 2008). However, this enabled scientists and investigators to design candidate gene case-control

TABLE 4: Genomic profiling in chronic heart failure (Benjamin and Schneider 2005)

COMPARISON	SUBJECTS	PLATFORM	FINDINGS
Failing versus non-failing	2 cases (1 ICM and 1 DCM) 2 control cases	Affymetrix Hu 6800	Alterations of expression of cytoskeletal and myofibrillar genes, genes encoding stress proteins, and genes involved in metabolism, protein synthesis, and protein degradation
Failing versus non-failing	7 cases (DCM) 5 control cases	Cardiochip (custom array)	Upregulation of genes for atrial natriuretic peptide, sarcomeric and cytoskeletal proteins, stress proteins, and transcription/translation regulators Down-regulation of genes regulating calcium signaling pathways
Failing versus non-failing	8 cases (DCM) 7 control cases	Affymetrix Hu 6800	103 differentially expressed genes with most prominent being airial natriuretic factor and brain natriuretic peptide
Failing versus non-failing	10 cases (DCM) 4 control cases	Custom arrays	364 differentially expressed genes Up-regulation being most prominent in genes for energy pathways, muscle contraction, electron transport, and intracellular signaling Down-regulation was most prominent in genes for cell cycle control
Failing versus non-failing	9 cases (5 ICM and 4 DCM) 1 control cases	Affymetrix HG-U95A	95 differentially expressed genes with notable up-regulation of atrial natriuretic peptide and brain natriuretic peptide Prominent pathways up-regulated include cell signaling and muscle contraction

Failing versus non-failing	6 cases (DCM) 5 control cases	Affymetrix HG-U133A	165 differentially expressed genes, the most prominent being structural and metabolic genes
Failing versus non-failing	5 cases (DCM) 5 control cases	Custom array for apoptotic pathways	Differentially expressed genes in apoptotic pathways
Pre- and post-left ventricular assist device	6 cases (3 DCM and 3 ICM)	Affymetrix Hu 6800	530 differentially expressed genes (295 up and 235 down) with prominent changes in genes for metabolism
Pre- and post-left ventricular assist device	7 cases (DCM)	Affymetrix HG-U133A	179 differentially expressed genes (130 up and 49 down) There was prominent up-regulation in nitric oxide pathways and down-regulation of inflamatory genes
Pre- and post-left ventricular assist device	19 cases (8 DCM and 11 ICM)	Affymetrix HG-U133A	107 differentially regulated genes (85 up and 22 down) Prominent was up-regulation of genes regulating vascular networks and down-regulation of genes regulating myocyte hypertrophy
HCM and DCM versus non-failing	3 DCM 2 HCM 3 control cases	Cardiochip (custom array)	Multiple genes and pathways up- and down-regulated some common to DCM and HCM some distinct to each

DCM, dilated cardiomyopathy; HCM, hypertrophic cardiomyopathy; ICM, ischemic cardiomyopathy studies (CGCS) in different complex conditions including HF (Vasan et al. 2009). A large CGCS in advanced HF employed 30,000 SNPs encompassing 2,000 candidate genes with a high *a priori* probability of cardiovascular involvement (Cappola, Li et al. 2010). This study reports identified and replicated two common genetic variants (rs1738943 and rs6787362) that are significantly associated with advanced HF in Caucasians. A similar approach led to the identification of inflammatory genes associated with DCM (Noutsias, Pankuweit et al. 2009; Jefferies and Towbin 2010). These and many other studies provide specific genetic data supporting the observation that the complex and heterogeneous syndrome of clinical HF results, in part, from an inherited predisposition (Lin, Gyenai et al. 2006).

The most important implication of candidate gene case-control study in HF is the identification of specific genes that may contribute to HF pathogenesis in humans. The study by Cappola, Li et al. (2010) highlights rs1738943 (combined $P = 3.09 \times 10^{-6}$), which is located in a 5-μ□ intronic region of the gene encoding the heat shock protein B7 (*HSPB7*), also known as the cardiovascular heat shock protein (cvHSP), as a member of the small heat shock protein family and is expressed almost exclusively in cardiac and skeletal muscles. An important function of *HSPB7* is to preserve contractile integrity by binding and stabilizing sarcomeric proteins (Meijering, Zhang et al. 2012). It is likely that variations in *HSPB7* may contribute to adverse cardiac remodeling. Mutations in another heat shock protein, *CRYAB* (also known as *HSPB5*) cause a rare form of cardiomyopathy (Hu, Yang et al. 2008). Further close scrutiny for rs1738943 points to its position high in the block of SNPs with high linkage disequilibrium that spans *HSPB7* and a nearby gene *CLCNKA*, which encodes a voltage-sensitive chloride channel protein expressed in the kidney (Barlassina, Dal Fiume et al. 2007). Variation in *CLCNKA* has been associated with alterations in renal sodium reabsorption and salt-sensitive hypertension (Sile, Vanoye et al. 2006). It is, therefore, likely that rs1738943 and other SNPs in the LD block spanning *HSPB7* and *CLCNKA* could be important in the pathogenesis of HF.

The other SNP rs6787362 associated with HF (combined $P = 6.09 \times 10^{-6}$) is located in a 3-μ□ intronic region of the *FRMD4B* gene. This gene encodes for FERM-domain-containing protein 4B (*FERMD4B*) that physically interacts with CYTH3, a downstream effector of PI-3 kinase signaling. PI-3 kinase is a mediator of many different signaling pathways, and it is difficult to speculate a specific mechanism based on the available data. It is believed that this protein may contribute to HF risk by conferring risk for coronary artery disease, while *HSPB7* is associated for both ischemic and non-ischemic subtypes (Cappola, Li et al. 2010). Further evidence may support the observation for both *HSPB7* and *FRMD4B* being novel susceptibility loci for clinical HF.

7.2   GENE-EXPRESSION (TRANSCRIPTIONAL) PROFILING

DCM is the endpoint of cumulative pathological changes in both ischemic (ICM) and non-ischemic cardiomyopathies (NICM). HF is the natural clinical consequence in both ICM and

NICM (Hare, Walford et al. 1992; Hunt, Baker et al. 2001). Apart from infiltration, inflammation, and volume overload, inherited factors account for a significant proportion of NICM. Among many hypotheses, it is commonly believed that there should be a common molecular pathway for inciting mechanisms that drives HF progression (Towbin and Bowles 2000). Although NICM and ICM have similar presentations (Hare, Walford et al. 1992), understanding their different pathophysiological mechanisms is essential in guiding future management.

The emergence of microarray technology to simultaneously assess mRNA levels of several thousand genes offers a novel approach to compare and contrast a number of different transcriptomes such as cancer (Singhal, Miller et al. 2008) and myocardium (Barrans, Stamatiou et al. 2001). It is likely that NICM and ICM share distinct and differentially expressed genes relative to normal hearts (Boheler, Volkova et al. 2003). In one experiment, differential gene expression patterns were observed by microarray analysis (Affymetric U133A GeneChips) in 21 NICM, and 10 ICM samples were compared with 6 nonfailing (NF) hearts (Kittleson and Hare 2005). Compared with NF hearts, 257 genes were differentially expressed in NICM and 72 genes in ICM. This approach was based on the group's previous study to determine clinical prediction algorithm using gene-expression biomarkers (Kittleson, Ye et al. 2004). However, 41 genes were shared between the two groups, mainly involving cell growth and signal transduction. Genes involving metabolism were frequent in NICM, and those in ICM had catalytic activity. This approach revealed some novel genes including angiotensin-converting enzyme-2 (ACE2), which was upregulated in NICM and not in ICM, demonstrating the different signaling pathways involved in HF pathophysiology. In contrast, significant downregulation of a member of the tumor necrosis factor (TNF) receptor subfamily (TNFRSF11B) was also observed in both NICM and ICM. This would probably account for increased TNF levels in HF correlated to disease severity (Rauchhaus et al., 2000). However, in clinical trials, the soluble TNF-α antagonist did not reduce the mortality in HF patients. This might be related to the downregulation of the TNF receptor in CHF.

Despite promising results of the gene-expression profiling in CHF using oligonucleotide microarrays, the technique has some limitations. Not all genes are represented on the Affymetric U133A arrays (Kittleson and Hare 2005). Therefore, so far, the knowledge acquired remains incomplete to warrant further studies. Nevertheless, the study by Kittleson, Minhas et al. (2005) offers novel insights into the unique disease-specific gene expression that exists between the end-stage cardiomyopathies of different etiologies. This approach illustrates that transcriptome analysis offers insight into pathogenesis-based therapies in HF management. Thus, in the future, gene-expression profiling may have a role in the precise diagnosis and instituting personalized therapeutic intervention.

7.3   EPIGENETICS AND EPIGENOMICS OF HEART FAILURE

There are several segments of DNA and RNA sequences across a genome that may have some functional significance in gene expression, modification, or control. Some of these carry evolutionary importance as shared by other biological creatures. The term epigenetics or epigenomics denote such sequences or a collective focal point lying in close proximity to the promoter region of a particular gene or a block of functionally relevant genomic polymorphisms. Mutations or structural changes in such genes or regions of genome could influence gene regulation or expression manifesting with the phenotype indistinguishable from mutations or structural changes within the gene itself. These changes are also parent-of-origin specific, thus providing the biological basis of genetic or genomic imprinting. There are now several recognizable clinically relevant conditions that are regularly encountered in the clinical genetic practice. In addition, it is firmly believed that epigenetic or epigenomic changes could also be important in the pathophysiology of several complex disorders and probably relevant to the outcome of pharmacotherapy, in particular, in chemotherapy for a whole range of cancer (Esteller 2008). It is likely that such genetic and genomic factors exist relevant to CHF.

Epigenomics has emerged as one of the most promising areas that will address some of the gaps in our current knowledge of the interaction between nature and nurture in the development of complex cardiovascular disease (CVD) such as CHF (Ordovas and Smith 2010) and cancer (Rodenhiser 2009). Several different epigenetic mechanisms are now identified notably DNA methylation, histone modification, and microRNA alterations, which collectively enable the cell to respond quickly to environmental changes (van Vliet, Oates et al. 2007). A number of environmental CVD risk factors, such as nutrition, smoking, pollution, stress, and the circadian rhythm, have been associated with the modification of epigenetic marks (Baccarelli, Rienstra et al. 2010). Epigenetic regulation differs from the more familiar genomic mechanisms, the primary targets of which include transcription factors that interact with DNA leading to several mechanisms including alternative splicing that allows synthesis of different protein isoforms by rearranging the information encoded in the exons of a specific gene. Epigenetic mechanisms modify the later steps in proliferative signaling pathways, including methylation of cytosine in genomic DNA, histone acetylation, and inhibition of RNA translation by small RNA sequences, called *microRNAs*. Cytosine methylation has been implicated in some familial cardiomyopathies (Robertson 2005) and histone acetylation can modify overload-induced cardiac hypertrophy (Backs and Olson 2006; Trivedi, Luo et al. 2007). Evidence that microRNAs regulate cardiac hypertrophy (Chien 2007; van Rooij, Sutherland et al. 2007) is of potential therapeutic importance because short RNA segments, called small-interfering (si)RNAs, can silence specific genes. The ability of siRNAs, which are readily synthesized commercially, to block specific proliferative pathways, promises additional approaches to inhibiting maladaptive hypertrophy to slow deterioration of failing hearts (Katz 2008).

Epigenomics represents a critical link between genomic coding and phenotype expression that is influenced by both underlying genetic and environmental factors (Figure 9). Available data indicate that cardiovascular risk factors might influence and remodel epigenomic patterns and that

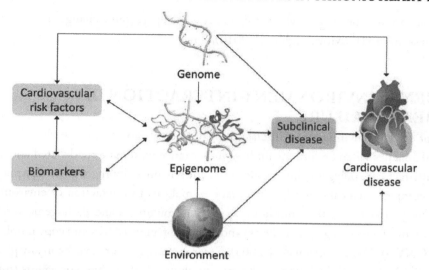

FIGURE 9: Interrelationship of environmental cardiovascular risk factors, biomarkers, genomic factors, and the epigenomic factors in the pathophysiology of a complex cardiovascular disease.

cardiovascular biomarkers are associated with epigenetic modifications. Epigenetic modifications may contribute to subclinical and clinical cardiovascular disease. Epigenomics is inherently inter-connected with genetics because epigenetic modifications can alter the expression of genetic varia-tions. Genetic variation is one of the determinants of DNA methylation and histone modifications. Epigenetic mechanisms, such as microRNA and histone modification, are crucially responsible for dysregulated gene expression in HF. In contrast, the role of DNA methylation, another well-characterized epigenetic mechanism, is unknown. It is argued that human cardiomyopathy of different etiologies is connected by a unifying pattern of DNA methylation pattern. Movassagh et al. (2010) performed genomic profiling in ischemic and idiopathic end-stage cardiomyopathic LV explants from patients who had undergone cardiac transplantation compared to normal control and identified three angiogenesis-related genetic loci that were differentially methylated. The techniques employed included sequencing using methylated-DNA immunoprecipitation-chip (MeDIP-chip), validating differential methylation loci by bisulfite-(BS) PCR, and high throughput sequencing. Using quantitative RT-PCR, the group found that the expression of these genes differed signifi-cantly between the CM hearts and the normal control ($p < 0.01$). Moreover, for each individual LV tissue, differential methylation showed a predicted correlation to differential expression of the cor-responding gene. Thus, differential DNA methylation exists in human cardiomyopathy compared to other systems, changes in DNA methylation at specific genomic loci usually precede changes in the expression of corresponding genes. These important observations in cardiomyopathy merit further investigation to determine whether epigenetic and epigenomic changes play a causative role in the progression of HF (Movassagh et al. 2010).

7.4   GENE–ENVIRONMENT INTERACTION IN HEART FAILURE

The concept of gene–environmental interaction is not new. It goes back to earlier times of human and medical genetics when definitions for basic inheritance patterns were debated and agreed. The basis of multi-factorial/polygenic inheritance rests entirely on assigning a particular phenotype, for example, a complex medical condition like diabetes mellitus to interaction of environmental (acquired or lifestyle) factors with a multitude of genetic (mutations and pathogenic variants within several hundreds of low-risk genes or alleles) and genomic factors (SNPs and copy number sequence variations (CNVs)) (Danziger, You et al. 2005). Environmental factors may be in any physical (heat, cold, irradiation), chemical (industrial pollutants, smoking, alcohol), biologic agents (microbes and parasites), or even emotional stress-triggering excessive neurohumoral sympathetic discharge. The role of wild genetic and genomic factors is to determine the biologic threshold on which the load of environmental factors would challenge the individual person's body organ or system. Thus, mutations or polymorphisms of any kind may lower the biologic threshold predisposing the individual to adverse consequences of environmental factors.

There is a complex interaction of the environment with the genetic and genomic loci (master switches) and regulatory elements (peptides and non-coding RNAs) (Ashrafian and Watkins 2007). These work together through a complex network of proteins (proteome) and transcriptional factors (transcriptome) resulting in variable complex phenotypes (Figure 10). The ultimate determinant of biologic phenotype is the pattern of cellular protein expression (i.e., the proteome). The proteome is, in turn, determined largely by transcription within the cell (i.e., the transcriptome). Microarray technology permits a comprehensive and quantitative description of the transcriptome. Traditional approaches to understanding disease pathogenesis have attempted to correlate genetic variants with gross phenotype. Instead, genetic correlations can be made with messenger RNA expression patterns as measured by arrays. These intermediate phenotypes are termed "expression quantitative traits" (eQTs). The combination of the systematic power of arrays to measure transcription and the power of genetic analysis to identify causality has proved to be powerful. The genetic determinants of gene expression, termed "expression quantitative trait loci" (eQTL), represent *cis* and *trans* regulatory elements in which variations cause alterations in cells' gene expression patterns. Identification of variants in these regulatory elements, termed "master switches," not only identifies key molecular protagonists in health and disease per se but also implicates downstream pathways with therapeutic implications.

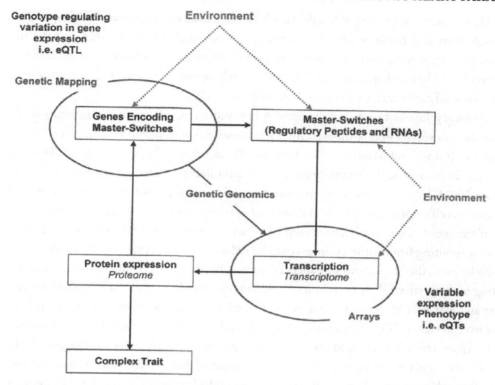

FIGURE 10: Complex interactions of environmental, genetic, and genomic factors in a biologic phenotype similar to chronic heart failure (Ashrafian and Watkins 2007).

7.5 MOLECULAR BASIS OF CHEMOTHERAPY-INDUCED HEART FAILURE

Cancer therapy has remarkably improved in recent years. Managing toxic effects of any chemotherapy regimen is a huge challenge for oncologists and other specialist clinicians. Cardiac toxicity is a major complication of cancer therapy, notably in managing leukemias and lymphomas and some other solid tumors (Monsuez, Charniot et al. 2010). Direct cardiac myocyte damage leading to chemotherapy-induced clinical cardiomyopathy is the basic underlying pathophysiologic mechanism that inevitably leads to CHF (Plana 2011). Two broad categories of chemotherapy agents are implicated in congestive HF complicating cancer therapy (Minotti, Salvatorelli et al. 2010). These include anthracyclines (Doxorubicin and Epirubicin) and tyrosine kinase inhibitors (Imitinab, transtuzumab, etc.).

Anthracyclines are a class of highly potent antitumor antibiotics utilized against hematologic and solid tumors in children and in adults (Grenier and Lipshultz 1998). Their use has been limited primarily by their cardiotoxic side effects, which may lead to congestive HF (Lipshultz 2006). Although there is a linear relationship between the cumulative dose received and the incidence of cardiotoxicity, in some patients, cardiotoxicity may develop at doses below the generally accepted threshold level (Kakadekar, Sandor et al. 1997). Early detection is essential in managing anthracycline-induced cardiotoxicity (Sandor, Puterman et al. 1992).

Anthracycline-induced cardiotoxicity is believed to be related to the generation of highly reactive oxygen species acting through lipid peroxidation, cause direct damage to cardiac myocyte membranes (Olson and Mushlin 1990; Jung and Reszka 2001; Zsáry, Szûcs et al. 2004). Another important factor may be the relatively poor antioxidant defense system of the heart (Takemura and Fujiwara 2007). In an attempt to circumvent these toxic effects, a wide variety of antioxidants have been used in cell culture, animal, and human studies without consistent beneficial effects. Moreover, none of the agents used, to date, are designed to act selectively upon the heart. If the cardiac complications resulting from anthracyclines could be reduced and/or prevented, higher doses could potentially be used, thereby increasing cancer cure rates. Furthermore, the incidence of cardiac toxicity resulting in congestive HF or even heart transplantation would be reduced, therefore increasing the quality and extent of life for cancer survivors. Several new and innovative approaches including targeting oxygen-derived free radicals and using antioxidant proteins are employed to limiting and/or preventing anthracycline-induced cardiotoxicity (Sparano 1998; Maradia and Guglin 2009).

Tyrosine kinases are a group of intracellular proteins that regulate cell division and cell proliferation through a cascade of several inter- and cross-linked protein kinases (PKs) (Alonso, Sasin et al. 2004). Several distinct clinical conditions are now recognized that result from activating mutations in genes encoding PKs including P13K, ERK, Ras, Raf, MAPK, etc. (Figure 11; Manning, Whyte et al. 2002). Collectively, P13K/RAS/MAPK protein cascade is dependent upon the regulatory control of several epidermal growth factor receptor (EGFR) family proteins (HER1, HER2, HER3, and HER4) with a broad domain extending from extracellular to intracytoplasmic region across the cell membrane (Tonks 2006). Activation or inhibition of HER protein molecules influences the Pi3K/RAS/MAPK cascade with clinical consequences including breast and colorectal cancers (Porębska, Harłozińska et al. 2000). Clinically recognizable multi-organ and multi-system conditions are delineated belonging to this major class of genetic conditions, for example, Costello syndrome and type 1 neurofibromatosis (Kim and Choi 2010). Most of these conditions involve tumorigenesis and cardiac complications including cardiomyopathy. Thus, it is not surprising that tyrosine kinase inhibitors (Imitinab and others) employed in cancer therapy may result in cardiotoxicity (Tables 5 and 6).

In this context, the development of a humanized monoclonal antibody against the HER2 protein (trastuzumab or Herceptin) as adjuvant therapy for early HER2-positive breast cancer ranks as one of the most satisfying and powerful examples of breast cancer therapy (Chien 2006). The

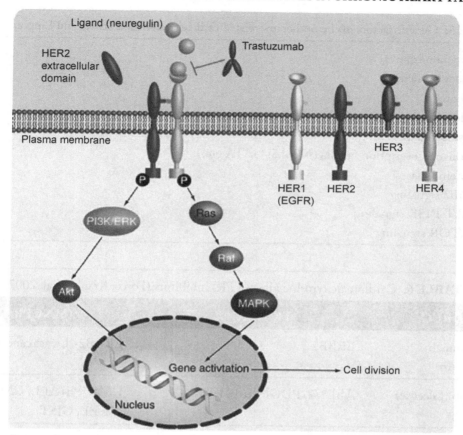

FIGURE 11: HER2 protein inhibition by transtuzumab regulating the P13K/RAS/MAPK proteins cascade (Force, Krause et al. 2007).

development of this antibody built on the discovery of the amplification of the HER2/neu (ErbB2) gene as a pivotal modifier in a subgroup of breast cancers (Piccart-Gebhart, Procter et al. 2005). A series of large-scale studies have conclusively shown that trastuzumab can substantially reduce the risk of recurrence and early death in women with HER2-positive breast cancer. However, this regimen is complicated with HF in 1% to 4% of patients, and 10% of patients may develop symptoms related to cardiac dysfunction (Vogel, Cobleigh et al. 2002). It is important to note that the incidence of cardiotoxic effects of trastuzumab appears to increase with exposure to anthracyclines, which mediate cardiac failure in a direct, dose-dependent manner.

In short, it seems as if trastuzumab both reduces the risk of recurrence of breast cancer and increases the susceptibility to HF (Smith, Procter et al.). A long-standing question is whether it is

---

TABLE 5: Genetic factors in chemotherapy-related cardiac dysfunction (Ewer and Lippman 2005)

- ➤ Mitochondrial
- ➤ Myocyte calcium
- ➤ Sarcomere
- ➤ Cytoskeletal
- ➤ Desmosomal
- ➤ Cardiac transcription factors (Nkx2.5; GATA etc.)
- ➤ Protein kinases
  - * Erb2 signaling
  - * AkT-P13K signaling
  - * mTOR signaling

TABLE 6: Cardiotoxic tyrosine kinase (TK) inhibitors (Force, Krause et al. 2007)

DRUG	TK TARGET	CANCER
Trastuzumab (Herceptin)	ERBB2	ERBB2+ breast cancer
Imatinib (Gleevac)	ABL1/2, PDGFRα/β, KIT	CML, Ph+ALL, CMML, CEL, GIST
Sunitinib (Sutent)	VEGFR1-3,KIT,PDGFRα/β, RET,CSF1R,FLT3	Renal cell Ca, GIST
Sorafenib (Nexavar)	VEGFR1-3, PDGFRβ, KIT, FLT3, RAF1, BRAF	Renal cell Ca, Melanoma
Lapatinib (Tykerb)	EGFR, ERBB2	Breast cancer
Cetuximab (Erbitux)	EGFR	Colorectal Ca, squamous cell Ca head/neck tumors

possible to delineate the biologic pathways that link trastuzumab to the onset of cardiotoxic effects, so as to reveal approaches to the design of drugs that would dissociate the beneficial effects from the adverse ones. Studies of mutant mouse models have documented a pivotal role of the *erbB2* gene in the embryonic and postnatal heart (Force, Krause et al. 2007). The induction of cardiac stress pathways by either hemodynamic overload or the cardiotoxicity of anthracyclines promotes the onset of

LV dysfunction in mice that are deficient in ErbB2 protein. These studies support a two-hit model of trastuzumab cardiotoxicity, in which there is a loss of ErbB2-mediated pathways that normally blunt the effects of stress-signaling pathways that anthracyclines activate in the heart. These results and findings in other relevant mouse models have suggested mitogen-activated protein kinase signaling as the triggering factor for the cardiotoxicity of trastuzumab (Rose, Force et al. 2010). Its basis is the fundamental role of ErbB2–ErbB4 heterodimeric receptors in triggering the myocyte-survival pathways that are required during the activation of acute stress signals. The loss of these survival cues after trastuzumab treatment can lead to the irreversible loss of cardiac myocytes during exposure to chemotherapeutic agents, such as the anthracyclines, which mediate overt heart dysfunction (Chien 2000). This reasoning is consistent with the clinical finding that trastuzumab also increases the risk of cardiac side effects in patients with existing forms of heart disease in which the cardiac stress signals are presumably already activated. However, direct proof of this concept has been elusive—it would require the dissociation of exposure to trastuzumab from cardiotoxic effects in humans.

The mechanism of this advance in the treatment of breast cancer might have been influenced by the doses and durations of trastuzumab therapy and anthracycline therapy, but the most likely, the explanation for the elimination of cardiotoxicity is the avoidance of either the concomitant administration of trastuzumab and anthracycline or the use of trastuzumab after anthracycline. It seems, therefore, that the risk of HF associated with trastuzumab was reduced because cardiac stress signals had not been activated by anthracyclines. Alternatively, trastuzumab can be given in therapeutically active doses with negligible cardiac side effects when combined with other chemotherapeutic agents such as docetaxel or vinorelbine (Joensuu, Kellokumpu-Lehtinen et al. 2006). However, data does not exist whether similar outcomes might hold in larger numbers of patients or in women with a preexisting heart disease. Despite several opposing hypotheses, oncologists and cardiologists agree that chemotherapy-related cardiac dysfunction (CRCD) is a distinct and probably a preventable entity (Ewer and Lippman 2005).

7.6   MOLECULAR BASIS OF ALLOGRAFT AND CARDIAC TRANSPLANT REJECTION

Cardiac transplantation is the last resort in the management of untreatable cardiac decompensation. Like in any other organ transplantation, allograft rejection is a major challenge in successful cardiac transplantation (Bieber, Stinson et al. 1970). Despite vigorous histocompatibility match, up to 10% cardiac allograft may be rejected (Michaels, Espejo et al. 2003). Histological changes of cardiac allograft rejection may include either myocardial disarray or coronary vasculopathy (Brunner-La Rocca, Schneider et al. 1998; Valantine 2004). The goal of post-cardiac transplantation care is early diagnosis of rejection and minimizing the side effects of the immunosuppressive therapy (Taylor, Barr et al. 1999; Horwitz, Tsai et al. 2004). Despite being an obvious invasive approach, the endomyocardial biopsy (EMB) remains the gold standard for diagnosing allograft rejection (Tan, Baldwin et al. 2007). There are several non-invasive methods available (echocardiography, ultrasonic myocardial backscatter, radionuclide imaging, magnetic resonance imaging, intra-myocardial electrograms, and multiparametric immune monitoring) but with limited sensitivity and often technically too difficult to validate and implement (Desruennes, Corcos et al. 1988; Hosenpud 1992).

Among many early host responses, circulating peripheral blood mononuclear cells (PMBCs) have attracted attention as a potential tool for early diagnosis of cardiac allograft rejection (Deng, Eisen et al. 2006). Measurement of PBMC gene expression might provide useful diagnostic information and reduce the need for EMB in patients who are asymptomatic. Recent studies using microarray analysis (Horwitz, Tsai et al. 2004) or real-time PCR analysis of cytokine genes (Schoels, Dengler et al. 2004) have suggested that gene expression measurements in PBMC may be correlated with cardiac allograft rejection. However, these studies are limited by the absence of methodology to recognize the imperfect "gold standard" nature of EMB, which creates significant challenges for diagnostic development and validation study design and analysis (Walter, Irwig et al. 1999). Based on the assumption that a gene expression signature of immune activation and leukocyte trafficking would be detectable in recipient PBMC and reflect the rejection status of the donor allograft (Deng, Eisen et al. 2006), it was hypothesized that a gene expression test could discriminate ISHLT grade 0 rejection (quiescence) from moderate/severe (ISHLT grade $\geq$ 3A) rejection (nonquiescence). This work was carried out at eight centers within the Cardiac Allograft Rejection Gene Expression Observational (CARGO) study (Deng, Eisen et al. 2006). Enrolled patients were followed at each clinical encounter with data collection including EMB, hemodynamics and/or echocardiography, immunosuppression, laboratory data and complications, which were captured in electronic clinical report forms. EMB slides were obtained from centers for interpretation by a panel of pathologists blinded to the clinical data. The study was conducted in three phases (Figure 12): (1) *candidate gene discovery* using a combination of focused genomic and knowledge-base approaches; (2) *diagnostic development* using PCR assays and rigorous statistical methods, and (3) *validation* in a prospective and blinded study. Samples were selected and divided into a training set, used for candidate gene discovery and diagnostic development, and a set used for validation of the gene expression signature described below. Despite a limited number of studies reported, gene expression profiling could be a promising tool in assessing the cardiac transplantation allograft rejection in the peripheral blood (Pham, Teuteberg et al.). This approach has potential applications of management following cardiac transplantation.

Apart from the cardiac transplant rejection, cases in which a male patient receives a heart from a female donor provide an unusual opportunity to test whether primitive cells translocate from the recipient to the graft and whether cells with the phenotypic characteristics of those of the recipient ultimately reside in the donor heart. The Y chromosome can be used to detect migrated

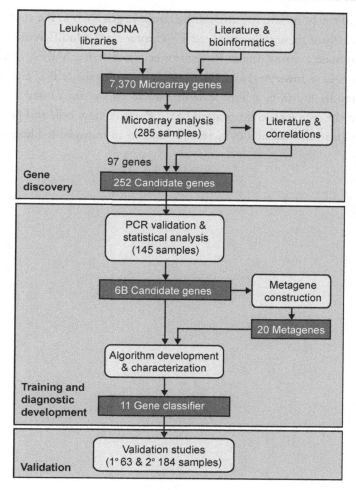

FIGURE 12: Noninvasive detection of cardiac allograft rejection (Deng, Eisen et al. 2006).

undifferentiated cells expressing stem cell antigens and to discriminate between primitive cells derived from the recipient and those derived from the donor (Quaini, Urbanek et al. 2002). This group examined samples from the atria of the recipient and the atria and ventricles of the graft by fluorescence in situ hybridization to determine whether Y chromosomes were present in eight hearts from female donors implanted into male patients. Primitive cells bearing Y chromosomes that expressed c-kit, MDR1, and Sca-1 were also investigated. Highly proliferative myocytes, coronary arterioles, and capillaries with the Y chromosome made up 7% to 10% in the donor hearts. Compared with the ventricles of control hearts, the ventricles of the transplanted hearts had markedly increased the numbers of cells that were positive for c-kit, MDR1, or Sca-1. The number of primitive cells was higher in the atria of the hosts and the atria of the donor hearts than in the ventricles of the donor hearts, and 12% to 16% of these cells contained a Y chromosome. Undifferentiated cells were negative for markers of bone marrow origin. Progenitor cells expressing MEF2, GATA-4, and nestin (which identify the cells as myocytes) and Flk1 (which identifies the cells as endothelial cells) were identified. These results illustrate a high level of cardiac chimerism caused by the migration of primitive cells from the recipient to the grafted heart. Putative stem cells and progenitor cells were identified in control myocardium and in increased numbers in transplanted hearts.

. . . .

CHAPTER 8

Molecular Basis of Reclassification (Taxonomy) of Heart Failure

8.1   CONVENTIONAL CLASSIFICATION OF HEART FAILURE

The clinical syndrome of chronic heart failure (CHF) can be viewed from different perspectives (Felker, Adams Jr et al. 2003; Maron, Towbin et al. 2006; see Figure 13):

1. Ventricular dysfunction resulting from a number of myocardial (myocarditis, ischemic changes, cardiomyopathy) and cardiac electrophysiological disorders (atrial fibrillation, ventricular tachycardia, and arrhythmias) compromising ventricular function.
2. Pressure overload in both pulmonary and systemic circulation primarily due to pulmonary arterial and systemic hypertension, respectively.

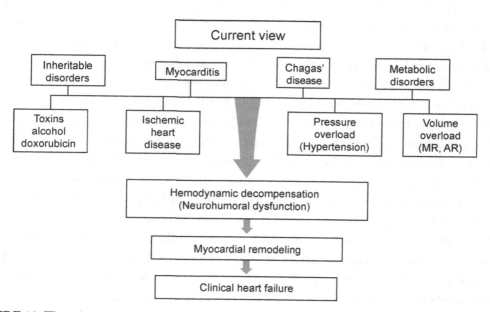

FIGURE 13: The schematic presentation of different causes and multiple adaptive pathways implicated in the clinical syndrome of heart failure (Cowie, Mosterd et al. 1997).

3.   Volume overload due to ventricular overfilling resulting from valvular dysfunction, for example, mitral regurgitation and atrial regurgitation.

All the above phases are followed by neurohumoral compensatory changes. However, in most cases, it is a short phase and leads to hemodynamic decompensation (Levine, Francis et al. 1983; Schrier and Abraham 1999). In addition, myocardial remodeling is another attempt to restore some ventricular function prior to establishment of the syndrome of clinical HF (Francis, McDonald et al. 1993; Cohn 1995).

8.2   NEW TAXONOMY FOR HEART FAILURE

Classification of a disease is a dynamic process reflecting the state of clinical and scientific knowledge and multi-faceted interpretations often influenced by sociocultural and environmental factors. A logical approach would require a common point from which different aspects of the disease could be assigned. In this context, the Darwinian term of taxonomy is coined. The Oxford Dictionary of the English language describes taxonomy as a systematic classification of organisms. In strict evolutionary terms, this could then be applied to the classification of disease based on genetic and genomic evidence validated for evolutionary biological significance (Bell 2004). In clinical medicine, commonly used etiologic categories (infection, inflammation, neoplasia, and aging) only indicate the final point of disease progression manifesting in defined symptoms and signs. Genetic and genomic factors are applicable in all these and other etiologic groups, thus strengthening the argument to accept genetic and genomic evidence fundamental to the origin of a disease (Kumar 2007).

In this article, evidence supporting genetic and genomic factors in the pathophysiology and clinical symptomatology of HF is highlighted. The conventional approach for the classification of HF is based on two major components of cardiodynamics—pressure and volume overload. Both can be independent or inter-dependent, one leading to another or vice versa acting in a cyclical manner. The other arbitrary approach is to distinguish between predominantly right or left-sided HF; that is, chronic stage may be inseparable when the clinical picture becomes biventricular HF. Most cardiologists, however, prefer to diagnose HF as predominantly systolic or diastolic, the latter being less well understood and evidently difficult to manage with often poor prognosis.

Despite differing approaches to classify HF, the basic pathology is progressive structural and functional changes in the myocardium, collectively put together under the broad umbrella term of cardiomyopathies (Maron, Towbin et al. 2006). Conventionally, myocardial damage is commonly ascribed to a number of acquired factors including coronary ischemia (commonly secondary to hyperlipidemia resulting in coronary atherosclerosis), infection (predominantly viral myocarditis), parasitic infestations (Chagas' disease), toxic (notably alcohol induced), and immunological process often triggered by the body's own immune system (for example, system lupus erythematoses). Whatever is the etiology, the cardiac decompensation goes through stages of LV hypertrophy (LVH), ventricular remodeling, and ventricular dilation (Figure 13). A similar process is also followed in classic Mendelian cardiomyopathies (HCM, DCM, and ARVC) (Hughes and McKenna 2005).

Studies on genetics and genomics of cardiomyopathies have helped in the reclassification of cardiomyopathies (Maron, Towbin et al. 2006). These studies have also revealed underlying genetic and genomic factors implicated in the pathophysiology related to apparent non-genetic causes of CHF (Creemers, Wilde et al. 2011). It is argued that subtle mutations and/or variants in HCM/DCM genes might be involved in small, but cumulatively discernable, contributions to HF predisposing the individual to one or more lifetime-acquired factors (Morita, Seidman et al. 2005). The new taxonomy for CHF reflects the complex interaction of environmental (acquired), genetic, and genomic factors (Ashrafian and Watkins 2007). Not surprisingly, this concept appears to be plausible in other common diseases to dissect out underlying molecular mechanisms and pathologic pathways in complex traits (Chen, Zhu et al. 2008; Goldstein 2009).

CHAPTER 9

Therapeutic Approaches in Chronic Heart Failure

The high incidence of HF has been met by several pharmacological, surgical, and therapeutic approaches (e.g., angiotensin inhibitors, pacemakers, and defibrillators) to treat and prevent early deaths owing to a specific heart condition (Hunt, Baker et al. 2001; American College of Cardiology Foundation/American Heart Association Task Force on Practice Guidelines, Heart et al. 2009). Some of these therapeutic strategies are modulated by underlying molecular defect or expressed genomic signatures (McLeod and Evans 2001; Donahue, Marchuk et al. 2006; Heidecker, Kasper et al. 2008; Table 2). An analysis of the available evidence reveals three broad therapeutic categories:

9.1   PHARMACOTHERAPY FOR CHRONIC HEART FAILURE

The pharmacotherapy for chronic heart failure aims to relieve symptoms, improve exercise tolerance, reduce incidence of acute exacerbations and reduce mortality. The key drugs to achieve these aims include ACE inhibitors and a beta-blocker. This combination is often aided by a diuretic in most patients to reduce symptoms of fluid overload.

An ACE inhibitor (for example *Ramipril, Captopril etc*) is generally prescribed to patients with asymptomatic left ventricular dysfunction (compensated) or symptomatic heart failure (decompensated). An angiotensin-II receptor antagonist (for example *Lasartan, Candesartan etc*) is now preferred to avoid side effects such as cough, and those who cannot tolerate ACE inhibitors.

Beta-blockers, preferably one of the cardio-selective types (for example *Bisoprolol or Carvediol*), are of value in any grade of stable heart failure and left-ventricular dysfunction. Beta-blocker treatment in heart failure should be started and supervised by experienced clinicians, at a very low dose and titrated very slowly over a period of weeks or months. Initial adverse response may require adjustment of concomitant therapy.

Patients with fluid overload are advised salt and fluid restriction and prescribed a loop or thiazide diuretic. A thiazide diuretic may be of benefit in patients with mild heart failure and good renal function. However, a thiazide diuretic is ineffective in patients with poor renal function (estimated creatinine clearance less than 30mL/minute) where a loop diuretic is recommended. It might be necessary to use both loop and thiazide diuretic to achieve optimal diuresis. It is important to closely monitor for fluid loss and electrolyte imbalance.

The aldosterone antagonist *spironolactone* is also used in the treatment of moderate to severe chronic heart failure in patients already taking an ACE inhibitor and a beta-blocker. A starting dose (less than 25 mg daily) may help reduce symptoms and mortality. Alternative aldosterone antagonists (for example *eplerenone*) are also used in severe heart failure with evidence of left ventricular dysfunction. It is important to monitor patients with regular serum creatinine and potassium measurements.

The use of digoxin is limited to patients with chronic heart failure with exercise intolerance and atrial fibrillation. It is also beneficial to patients with sinus rhythm who remain decompensated despite adequate combination therapy for heart failure (ACE inhibitor, beta-blocker and a diuretic). A combination of isosorbide dinitrate with hydralazine might help patients who remain symptomatic even with digoxin and standard heart failure regimen. However, Afro-American patients might not tolerate the combination of isosorbide dintrate and hydaralazine. Nevertheless this combination might help in addition to the standard heart failure therapy.

9.2   PERSONALIZED AND EVIDENCE-BASED DEVICE THERAPY IN HEART FAILURE

Left heart assist devices. Genomic expression profiling in patients with HF managed by left heart assist devices indicate certain significant biological phenomenon. The apoptosis study (Steenbergen, Afshari et al. 2003) report that patients with heart assist devices had distinct myocardial profiles. This led to a number of follow-up further studies on LV assist device implantation. In non-ischemic patients, molecular changes included downregulation of certain force transmission and neurohormonal genes and an enhancement of metabolic genes (Blaxall, Tschannen-Moran et al. 2003). Differential expression of apelin-related genes was observed in another study using 12,000-element microarray chip (Chen, Park et al. 2003). The apelin serves as a biomarker for early stage HF (Liew and Dzau 2004). Another study found that nitric oxide-related genes are activated in HF patients managed by LV assist devices (Chen, Park et al. 2003) as well as differential expression of vascular signaling genes and *GATA4* (a mediator of myocyte hypertrophy) (Hall, Grindle et al. 2004).

Cardiac resynchronization therapy. Another aspect of interest in the management of CHF is the use of cardiac resynchronization therapy (CRT) to bring synchronization of ventricular function. Cardiac electromechanical dyssynchrony causes regional disparities in workload, oxygen consumption, and myocardial perfusion within the left ventricle. It is argued that such dyssynchrony also induces region-specific alterations in the myocardial stress-response kinases and cytokines, more noted in the left ventricle (Freemantle, Tharmanathan et al. 2006; Cleland, Freemantle et al. 2008). It has been demonstrated that CRT can reverse this process (Chakir and Kass). This observation has led to further analysis of broader biological impact of CRT facilitating the electromechanical function (Barth, Aiba et al. 2009). This hypothesis has been shown to be applicable using a global gene expression profiling approach in a recently developed canine model of dyssynchronization HF and CRT, examining regional disparities in the cardiac transcriptome in dyssynchronous HF (DHF) and determining the capacity of CRT to ameliorate these abnormalities (Barth, Aiba et al. 2009). Dyssynchrony-induced gene expression changes in 1050 transcripts were reversed by CRT to levels of NF hearts (false discovery rate < 5%). CRT remodeled transcripts with metabolic and cell signaling function and greatly reduced heterogeneity of gene expression compared with DHF. Thus, DHF is not simply the severe HF but a form of disease with profound regional gene expression disparities. The animal experiments convincing evidence that by re-coordinating contraction, gene-expression heterogeneity can be returned to normal in a failing heart on a global genomic level. Further successful replications of these observations may point to a genomic method to assess the severity of HF and monitor the impact of CRT (Barth, Chakir et al.; Barth, Aiba et al. 2009).

9.3 PHARMACOGENETICS AND PHARMACOGENOMICS OF HEART FAILURE

Pharmacogenomics is a rapidly evolving new discipline in clinical pharmacology and drug therapeutics. Apart from utilizing the genomic and proteomic advances in drug discovery and new drug development, pharacogenomics offers a powerful tool to predict an individual's drug response and outcome (Liggett 2010). It is closely linked to pharmacogenetics, a term coined in the mid-1950s that highlighted few Mendelian genetic conditions as the basis of adverse drug response. For example, exaggerated hemolytic crisis in individuals with G6PD was treated with primaquine for malaria (Na Bangchang, Songsaeng et al. 1994). Specific prescribing guidance exists against several drugs in relation to a number of genetic disorders (Shah 2004; Gardiner and Begg 2005).

In the context of HF, a clinician is fully aware of variable pharmacologic response among individuals taking various drugs. There are two main factors—phenotypic heterogeneity of HF and individuals' genome-wide polymorphisms (Liggett 2010; Eichelbaum, Ingelman-Sundberg et al. 2006). Recent studies have demonstrated that non-synonymous SNPs of the β_1-adrenergic receptor (β_1-AR), a member of the seven membrane-spanning receptor superfamily, alters the therapeutic response to β-blockers during HF (Toda 2003; Liggett 2010) used microarrays to characterize alterations in gene expression induced by angiotensin II receptor blockers. Another study report attenuation of the HF risk by GSPE antioxidant action by inhibiting the *CD36* cardioregulatory gene (Bagchi, Sen et al. 2003).

9.4   TARGETED MOLECULAR THERAPY FOR HEART FAILURE

Specific therapeutic modalities for several acute and chronic diseases have been developed using the genetic and genomic knowledge and molecular understanding of underlying pathological mechanisms. This is also applicable to CHF (Table 7). There are potential therapeutic measures specifically targeted at the particular protein molecule acting in isolation or within a gene-molecular pathway. Some of these new drugs have crossed the phase 3 clinical trials and are licensed for clinical use.

TABLE 7: Potential therapeutic targets for clinical heart failure (Liew and Dzau 2004)		
AREA OF PATHOGENESIS/ GENE/PROTEIN	**THERAPEUTIC MEASURE**	**REFERENCES**
Calcium channel dysfunction	Normalizing calcium cycling	(Wehrens, Lehnart et al. 2005)
Ventricular remodeling	MMO modulation	(Li, McTiernan et al. 2000)
Cardiac contractility defects	Regulation of β-adrenergic receptor signaling	(Yasuda and Lew 1997)
Neurohormones and cytokine dysfunction	β-Adrenergic receptor blockers	(Keulenaer and Brutsaert 2007)
Neurohormones and cytokine dysfunction	ACE inhibitors and angiotensin II receptor blockers	(Manohar and Piña 2003)
Signal transduction dysfunction	Kinase inhibitors	(Brodde, Michel et al. 1995)
Hypertrophic growth	Calcineurin inhibitors and HDAC inhibitors	(Frey, Barrientos et al. 2004)
Apoptosis	Immunoadsorption of anti-β1-AR and anti-cardiac	(Segers and Lee 2008)
Autoimmune	Immunoadsorption of anti-β1-AR and anti-cardiac antibodies	(Kaya, Leib et al. 2012)

ACE—angiotensin converting enzyme; AR—adrenergic receptor; HDAC—histone deacetylase; MMP—matrix metalloproteinase; TNF—tumor necrosis factor

9.5   GENE THERAPY IN HEART FAILURE

Among many facets of medical genetics, gene therapy for inherited disorders remains the most elusive and challenging goal. Inherited cardiovascular conditions are no exception. With increasing knowledge of basic molecular mechanisms governing the development of HF, the possibility of specifically targeting key molecular targets is not impossible. Potential gene therapy approaches are being evaluated equipped with the technological knowledge of in vivo transduction of myocardial tissue with long-term expression of a transgenic animal model. However, gene therapy in HF is in its infancy. Nevertheless, this challenging and promising field is gaining momentum with few phase 1 clinical trials for HF gene therapy (Vinge, Raake et al. 2008).

Gene therapy for CHF requires success in two parameters, first, to clearly identify the beneficial or detrimental molecular targets and second, to develop sound technical capability to manipulate these targets at the molecular level in sufficiently a large number of cardiac cells. However, several obstacles are encountered in achieving safe and efficient gene transfer to human myocardium. In broad terms, there are two distinct aspects of gene therapy in HF-gene therapy techniques and potential molecular targets. Gene therapy, in principle, includes two distinct components—technical and molecular manipulations for achieving the desired effect. Technical approaches are essentially similar to any gene therapy application.

HF gene therapy technical approaches. Gene therapy for HF or any other pathological state must be aimed at correcting the key molecular mechanisms that reduce/reverse the inevitable deterioration of the physiological function. In the case of an inherited cardiomyopathy, this requires introduction of DNA/RNA that targets specific cardiomyocyte processes involved in HF outcomes, for example, lethal arrhythmias, acute cardiac decompensation, and prevention of end-stage pump failure. This is a challenging task and unlikely to be achieved by simple "one-target-fits-all" approach. However, a single-target approach may be applicable for HF as common biological and molecular signaling pathways are often involved. The ideal HF gene therapy strategy may depend on the specific cause of the condition (i.e., ischemic, valvular, hypertensive, or monogenic cardiac condition). A combinatorial approach may also be applicable that targets different cardiac cells and signaling pathways (Vinge, Raake et al. 2008). For example, HF gene therapy for IHD may ideally involve targeting vascular cells by stabilizing coronary plaques and inducing neoangiogenesis, providing means to prevent death of surviving cardiomyocytes (for example, by inhibiting apoptosis), preventing or reducing cardiac remodeling by targeting fibroblasts and targeting myocardial electrophysiological abnormalities for reducing risk of arrhythmias. These basic principles may also apply to other etiologic groups.

Genetic intervention in any gene therapy approach may be targeted to a number of molecular mechanisms and may include (a) overexpression of a target molecule, (b) alteration of the target's intracellular shuttling routes through decoy molecules, (c) loss-of-function approaches using dominant negative molecules or by the introduction of RNA interference, (d) correcting deleterious gene mutations/deletions at the genome or primary mRNA level, and (e) installing genetically modified donor cells (stem cells and/or differentiated cells) (Vinge, Raake et al. 2008). It is important to appreciate that the successful outcome of gene therapy would finally depend on the selection of vectors and gene delivery techniques.

Vehicles for transduction/transgenic expression. The selection of a suitable vehicle or vector is a major challenge for any gene therapy program. In the case of HF, this would require efficient myocardial transduction and long-term transgenic expression. So far, only viral vectors appear to meet such requirement. However, concerns remain for possible pathogenic effects of the viral vector and/or adverse outcomes of gene therapy. Nevertheless, viral vectors remain a preferred choice. Among many viral vectors, adenoviruses (AdVs) and adeno-associated viruses (AAVs) are preferred choices for HF gene therapy (Vinge, Raake et al. 2008).

In vivo experimental studies have demonstrated effective myocardial transduction with AdVs to alter global cardiac functions (Williams and Koch 2004). However, the major disadvantage of AdVs is an inflammatory response and secondary immune effect targeted at cardiac cells (Jooss and Chirmule). In contrast, AAVs are generally considered more promising for gene therapy of chronic diseases such as HF because they readily infect cardiac tissue, produce stable and long-term transgene expression, and are much less immunogenic (Williams and Koch 2004). So far, AAVs have not been shown to cause any known human disease (McCarty 2008). In addition, some serotypes of AAVs appear to stimulate cardiac cells, a feature that may be exploitable to improve HF gene therapy outcomes (Williams, Ranjzad et al.).

Gene delivery techniques. A successful gene therapy strategy would carefully plan for effective and efficient gene delivery technique. Ideally, the intravenous route is best suited for this purpose. However, this could be an unattainable goal in humans due to the large blood volume and subsequent dilution issues. However, this could be an advantage given cardiac tropism concerns for some AAVs serotypes as this may lessen the amount of virus needed for effective human cardiac gene transfer. So far, cardiac gene therapy strategies include either intravascular (through coronary arteries) or direct cardiac muscle approach. Intracoronary gene delivery is obviously clinically relevant and appealing; however, this approach is generally inefficient for myocardial gene transfer unless adjuvants (for example, vascular endothelial growth factor) and specific conditions (for example, increasing perfusion pressure) are used for enhancing contact time and, thus, facilitating transduction to cardiac cells (O'Donnell and Lewandowski 2005). In an attempt to further optimize intracoronary delivery, recirculation of the virus by a closed-loop system is also being developed (O'Donnell and Lewandowski 2005).

HF gene therapy targets. A number of molecular targets have been considered for gene therapy in HF. A detailed account of these approaches is beyond the scope of this review. Among these, cardiomyocyte sarcoplasmic reticulum calcium is probably the chief target that features in most HF gene therapy protocols (Vinge, Raake et al. 2008).

9.5.1 Cardiomyocyte Calcium Handling Proteins

There is considerable evidence that cardiomyocyte Ca^{2+} handling is important for excitation-contraction (EC) coupling. This important feature has long been the focus for HF molecular therapy and gene therapy. The EC coupling begins with the initiation of a cardiomyocyte action potential and Ca^{2+} enters the cell through voltage-gated L-type Ca^{2+} channels. The initial influx of Ca^{2+} triggers the ryonidine (RyR) to extrude Ca^{2+} from the sarcoplasmic reticulum (SR) into the cytosol (Bers 2008). This Ca^{2+}-induced Ca^{2+} release triggers cardiomyocyte contraction through Ca^{2+} binding to troponin C within the myofilaments of the sarcomere (Bers 2002). The calcium extrusion is equally important for the relaxation of the sarcomere facilitated by SERCA2a and sarcolemmal Na^+–Ca^{2+} exchanger (NCX). In HF, there is a decreased SR Ca^{2+} content and a prolonged Ca^{2+} transient, which is generally considered to be a consequence of increased NCX, a reduction of SERCA2a and a decreased PLN/SERCA2a ratio as well as "leakiness" of the RyR (Periasamy and Janssen 2008). In addition to these changes causing dysfunctional contractile performance, the decreased clearance of cytosolic Ca^{2+} may increase the risk of arrhythmias and may also precipitate pathological cardiac remodeling (Bers, Despa et al. 2006).

Several molecules involved in dysfunctional Ca^{2+} are attractive for HF gene therapy (Figure 6). Some of these selected include SR Ca^{2+-} ATPase (Hasenfuss et al., 1994), PLN (Pattison, Waggoner et al. 2008), protein phosphatase 1 and inhibitor protein-1 (Ceulemans and Bollen 2004), parvalbumin (Eberhard and Erne 1994), and S100A1 (Most, Remppis et al. 2007).

9.5.2 Other Potential Targets

Among the other potential HF gene therapy targets, the beta adrenergic signaling proteins deserve particular attention. Several members of the β-adrenergic signaling pathway have been examined with respect to their role in the maintenance of normal cardiac function as well as their influence on HF development. Two basic observations are worth reminding, first, the pharmacological use of β-AR antagonists in HF is generally beneficial, and second, sustained inotropic support in human HF through increased β-AR activation is detrimental (Wilk, Myers et al. 2006). In this context, genes encoding proteins for beta-1 AR downregulation and beta-2 AR overexpression are also potential targets.

Other potential HF gene therapy targets include *kinase signaling proteins* (for example, inhibition of GRK and AC type 6) and *dystrophin-Sarcoglycan complex proteins*. Detailed description and discussion on this is beyond the scope of this review.

· · · ·

CHAPTER 10

Summary and Future Directions

The syndrome of clinical HF is heterogeneous and complex. In-depth analysis of pathophysiology in major categories of HF reveals complex molecular pathways that carry unique genetic and genomic signatures. A small but clinically significant high-risk group of patients belong to the Mendelian cardiovascular conditions (HCM, DCM, and others; see Table 2) that form the bulk of inherited cardiovascular conditions with clinical HF as the major phenotype during and toward the end of the natural course of the disease. Mutations in gene-encoding proteins belonging to the sarcomere, cytoskeletal, desmosomal, and cytokine families are closely associated with this group of conditions. High-risk multi-generation families with autosomal dominant HCM may reveal pathogenic mutations in approximately two thirds of symptomatic individuals. However, missense mutations and variants of unknown significance are likely to be implicated in the medium- to low-risk group of patients with clinical HF, often sporadic or unconvincing limited family history. The phenotype in these patients may closely resemble a specific Mendelian condition but lacks sufficient power of discrimination. It is argued that these patients carry a combination of genetic or genomic signatures that probably function in the background of acquired factors. Errors in mRNA processing may also account for variable clinical phenotypes in some of these patients. In addition, post-translational modifications of proteins can significantly alter their function.

Despite differing mechanical factors (pressure or volume overload), the myocardium elicits a number of common responses at the level of gene expression. Gene expression of high- to medium-risk alleles is influenced by alternative splicing, DNA methylation, histone modification, and miRNA. In addition, the modifier effect of the polymorphic variant close to the promoter end or epigenetic changes might also influence gene expression (Chakravarti and Kapoor). However, these common responses cannot explain the well-known individual disease variability that occurs in HF. As individual genomes differ by millions of bases, it is quite likely that individual variability in the pathophysiology of HF depends not only on the underlying etiology but also on the genetic or genomic variants (Figure 13). In this context, the role of non-coding RNA, for example, miRNA, could be important in molecular and cellular mechanisms in HF. Sequence variation in miRNA target sequences may be an important source of phenotypic variability. The first example was the finding that a mutation in the 3′ UTR of the myostatin gene in sheep created an miRNA binding site, so that miRNA-mediated suppression of myostatin led to hypermuscularity (Schuelke, Wagner et al. 2004). A mutation creating a potential illegitimate microRNA target site in the myostatin gene affects muscularity in sheep (Chen and Rajewsky 2006). However, this mechanism has not yet been demonstrated in human heart disease.

Another recent example of individual variability was shown by variation in transcription factor binding among humans (Kasowski, Grubert et al.). Kasowski's group used ChIP-seq to map nuclear factor-κB (NF-κB) and RNA polymerase II (Pol II) binding sites in 10 human lymphoblastoid cell lines. Interestingly, as much as 25% of the Pol II binding regions and 7.5% of the NF-κB binding regions differed significantly between individuals. This greatly exceeds estimates for sequence variation in coding sequences (0.025% within humans) and suggests a strong role for variation in transcription factor binding to explain human diversity. Differences in the transcription factor or Pol II binding in this study were frequently associated with SNPs. Mapping variants that alter gene expression may help us to understand why individual responses to cardiac loading can vary. Gene expression profiling has a limited but significant place in molecular genetic investigations in Mendelian families with a variant of unknown significance related to one of the genes belonging to molecular pathway families. However, the technical expertise in clinically useful gene expression profiling method is restricted to few specialized laboratories. Sophisticated real time PCR (RT-PCR) method and Northern blot analysis are some of the techniques. This approach is also helped by the development of non-invasive "blood-cell derived RNA" technique that avoids the use of cardiac tissue samples (Whitney, Diehn et al. 2003).

Majority of patients with clinical HF often have an identifiable environmental cause such as ischemia, excessive alcohol consumption, infection, and chronic inflammation. However, despite similarities in gender, ethnicity, age, and lifestyle, the natural history and clinical outcome may vary tremendously. This is attributed to inter-individual genomic variation as evidenced by a number of genome-wide association studies (GWAS). There is overwhelming evidence to support the argument that genetic and acquired forms of HF induce a common cardiac response (Figure 14). This figure schematically depicts that the different etiologies underlying HF (genetic and acquired forms) all induce a mismatch between cardiac loading and capacity. This imbalance provokes a broad and rather common cardiac response. The precise nature and intensity of this response is expected to be crucial in determining disease severity. Individual variation in the genomic sequence of regulatory elements or 3′ UTRs, for example, can augment or inhibit this response and thereby cause considerable individual variability in the severity of the resulting disease. ECM, extracellular matrix; miRNA, microRNA.

Genomic signatures (SNPs, CNVs) along with the clinical phenotype could offer tremendous diagnostic discriminatory power and might help in the prediction of long-term outcomes. This is rapidly evolving with the availability of large microarray genomic platforms from 500K upward.

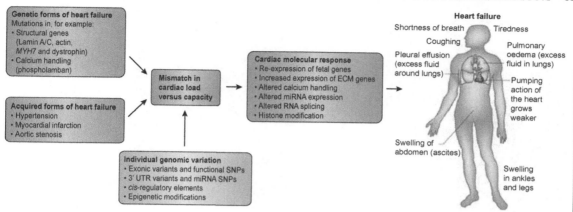

FIGURE 14: Close relationship of genetic, genomic, and acquired factors in the pathophysiology and clinical manifestations in chronic heart failure (adapted with permission from Creemers, Wilde et al. 2011).

The whole genome or selected genome analysis using microarray technology has the potential of clinical applications in a number of complex and heterogeneous conditions including the syndrome of clinical HF. In the future, clear genetic and genomic signatures for CHF are likely to be made available on the public domain facilitating the clinician in the management. In addition, this might also allow identification of new drug targets (see Table 7) and the ability to predict drug response in a particular phenotype.

Unbiased genomic technologies are sketching a picture of the various levels at which the myocardium responds to increasing volumes and pressures. Although uniform changes take place at every level of transcriptional and translational control, large-scale genetic studies provide the first steps for stratifying etiological and therapeutic subgroups within the broad syndrome of HF. Advances in mapping the inter-individual genetic variants are expected to have an important effect on individualized therapy. The cornerstones of HF treatment are β blockers and angiotensin-converting enzyme (ACE) inhibitors: variants in three different genes have already been shown to influence specific β-blocker treatment (β1-adrenergic receptor, G protein-coupled receptor kinase and cytochrome P450), and variation in genes related to ACE inhibition modifies treatment outcomes in patients with coronary artery disease. These are the first suggestions that HF treatment may be directed by pharmacogenetic approaches.

CHAPTER 11

Learning Outcomes

1. The clinical syndrome of CHF is vastly heterogeneous and results from progressive interaction of both genetic and acquired etiological factors.

2. Commonly recognized acquired factors include chronic inflammation (infective, parasitic, or immunological), acute or chronic recurrent ischemic myocardial damage, and mechanical myocardial damage and remodeling due to chronic pressure or volume overload (hypertension and aortic stenosis).

3. Genetic factors include pathogenic mutations or polymorphic variants within genes encoding many myocardial proteins, for example, sarcomere, desmosome, taffazin, and many more. These are collectively assigned as Mendelian cardiomyopathies manifesting with CHF.

4. Apart from single genes, structural changes, CNVs, and SNPs scattered across the genome also play an important role in the molecular pathophysiology of CHF. Precise mechanisms are unclear but could be through multitude modes including modifier or epigenetic functions.

5. As individual genomes differ by millions of bases, it is quite likely that individual variability in the pathophysiology of HF depends not only on the underlying etiology but also on genetic and genomic variants.

6. Despite tremendous clinical variation, the myocardium elicits a number of common responses at the level of gene expression, notably alternative splicing, DNA methylation, histone modification, and miRNA expression. However, these common responses cannot explain the well-known individual disease variability that occurs in HF.

7. Recent unraveling of miRNA mechanisms in the heart is of particular interest. Sequence variation in miRNA target sequences may be an important source of phenotypic variability.

8. Advances in mapping the inter-individual genetic variants are expected to have an important effect on individualized therapy. These are the first suggestions that HF treatment may be directed by pharmacogenetic approaches.

9. The cornerstones of HF treatment are β-blockers and angiotensin-converting enzyme (ACE) inhibitors or receptor blocker variants in three different genes have already been shown to influence specific β-blocker treatment (β1-adrenergic receptor, G protein-coupled receptor kinase and cytochrome P450) and variation in genes related to ACE inhibition modifies treatment outcomes in patients with coronary artery disease.

10. Unbiased next-generation genomic technologies are sketching a picture of the various levels at which the myocardium responds to increasing volumes and pressures. Variable pattern of myocardial changes probably reflect transcriptional and translational control of the gene expression.

11. Large-scale genetic studies provide the first steps for stratifying etiological and therapeutic subgroups within the broad syndrome of HF.

• • • •

References

Alcalai, R., J. G. Seidman, et al. (2008). Genetic basis of hypertrophic cardiomyopathy: from bench to the clinics. *Journal of Cardiovascular Electrophysiology* 19(1): pp. 104–10.

Aldred, M. A., J. Vijayakrishnan, et al. (2006). BMPR2 gene rearrangements account for a significant proportion of mutations in familial and idiopathic pulmonary arterial hypertension. *Human Mutation* 27(2): pp. 212–3.

Alonso, A., J. Sasin, et al. (2004). Protein tyrosine phosphatases in the human genome. *Cell* 117(6): pp. 699–711.

Ambros, V. (2001). microRNAs: Tiny regulators with great potential. *Cell* 107(7): pp. 823–6.

American College of Cardiology Foundation/American Heart Association Task Force on Practice Guidelines, I. S. F. Heart, et al. (2009). 2009 Focused Update: ACCF/AHA Guidelines for the diagnosis and management of heart failure in adults. *Journal of the American College of Cardiology* j.jacc.2008.11.009.

Anonymous. Current world literature. *Current Opinion in Cardiology* 25(3): pp. 282–97 10.1097/HCO.0b013e3283394f8b.

Arab, S. and P. P. Liu (2005). Heart failure in the post-genomics era: gene–environment interactions. *Current Opinion on Molecular Therapeutics* 2005 Dec;7(6): pp. 577–82.

Asakawa, M., H. Takano, et al. (2002). Peroxisome proliferator-activated receptor γ plays a critical role in inhibition of cardiac hypertrophy in vitro and in vivo. *Circulation* 105(10): pp. 1240–6.

Ashrafian, H. and H. Watkins (2007). Reviews of translational medicine and genomics in cardiovascular disease: new disease taxonomy and therapeutic implications: cardiomyopathies: therapeutics based on molecular phenotype. *Journal of the American College of Cardiology* 49(12): pp. 1251–64.

Baccarelli, A., M. Rienstra, et al. (2010). Cardiovascular epigenetics. *Circulation: Cardiovascular Genetics* 3(6): pp. 567–73.

Backs, J. and E. N. Olson (2006). Control of cardiac growth by histone acetylation/deacetylation. *Circulation Research* 98(1): pp. 15–24.

Bagchi, D., C. K. Sen, et al. (2003). Molecular mechanisms of cardioprotection by a novel grape seed proanthocyanidin extract. *Mutation Research/Fundamental and Molecular Mechanisms of Mutagenesis* 523–524(0): pp. 87–97.

Barlassina, C., C. Dal Fiume, et al. (2007). Common genetic variants and haplotypes in renal CLCNKA gene are associated to salt-sensitive hypertension. *Human Molecular Genetics* 16(13): pp. 1630–8.

Barrans, J. D., D. Stamatiou, et al. (2001). Construction of a human cardiovascular cDNA microarray: portrait of the failing heart. *Biochemical and Biophysical Research Communications* 280(4): pp. 964–9.

Barth, A., K. Chakir, et al. (2012). Transcriptome, proteome, and metabolome in dyssynchronous heart failure and CRT. *Journal of Cardiovascular Translational Research* 5(2): pp. 180–7.

Barth, A. S., T. Aiba, et al. (2009). Cardiac resynchronization therapy corrects dyssynchrony-induced regional gene expression changes on a genomic level/clinical perspective. *Circulation: Cardiovascular Genetics* 2(4): pp. 371–8.

Bell, J. (2004). Predicting disease using genomics. *Nature* 429(6990): pp. 453–6.

Benjamin, I. J. and M. D. Schneider (2005). Learning from failure: congestive heart failure in the postgenomic age. *Journal of Clinical Investigation.* 2005 Mar;115(3): pp. 495–9.

Bers, D. M. (2002). Calcium and cardiac rhythms. *Circulation Research* 90(1): pp. 14–7.

Bers, D. M. (2008). Calcium cycling and signaling in cardiac myocytes. *Annual Review of Physiology* 70(1): pp. 23–49.

Bers, D. M., S. Despa, et al. (2006). Regulation of Ca2+ and Na+ in normal and failing cardiac myocytes. *Annals of the New York Academy of Sciences* 1080(1): pp. 165–77.

Bhatia, R. S., J. V. Tu, et al. (2006). Outcome of heart failure with preserved ejection fraction in a population-based study. *New England Journal of Medicine* 355(3): pp. 260–9.

Bieber, C. P., E. B. Stinson, et al. (1970). Cardiac transplantation in man: VII. Cardiac allograft pathology. *Circulation* 41(5): pp. 753–72.

Blaxall, B. C., B. M. Tschannen-Moran, et al. (2003). Differential gene expression and genomic patient stratification following left ventricular assist device support. *Journal of the American College of Cardiology* 41(7): pp. 1096–106.

Blayney, L. M. and F. A. Lai (2009). Ryanodine receptor-mediated arrhythmias and sudden cardiac death. *Pharmacology & Therapeutics* 123(2): pp. 151–77.

Boheler, K. R., M. Volkova, et al. (2003). Sex- and age-dependent human transcriptome variability: implications for chronic heart failure. *Proceedings of the National Academy of Sciences* 100(5): pp. 2754–9.

Bonow, R. O., S. Bennett, et al. (2005). ACC/AHA clinical performance measures for adults with chronic heart failure. *Circulation* 112(12): pp. 1853–87.

Bracco, L., E. Throo, et al. (2006). Methods and platforms for the quantification of splice variants' expression. *Progress in Molecular and Subcellular Biology* 44: pp. 1–25.

Braunwald, E. (2008). The management of heart failure. *Circulation: Heart Failure* 1(1): pp. 58–62.

Brodde, O.-E., M. C. Michel, et al. (1995). Signal transduction mechanisms controlling cardiac contractility and their alterations in chronic heart failure. *Cardiovascular Research* 30(4): pp. 570–84.

Brunner-La Rocca, H. P., J. Schneider, et al. (1998). Cardiac allograft rejection late after transplantation is a risk factor for graft coronary artery disease. *Transplantation* 65(4): pp. 538–43.

Cappola, T. P., M. Li, et al. (2010). Common variants in HSPB7 and FRMD4B associated with advanced heart failure / clinical perspective. *Circulation: Cardiovascular Genetics* 3(2): pp. 147–54.

Ceulemans, H. and M. Bollen (2004). Functional diversity of protein phosphatase-1, a cellular economizer and reset button. *Physiological Reviews* 84(1): pp. 1–39.

Chaanine, A. H., J. Kalman, et al. (2010). Cardiac gene therapy. *Seminars in Thoracic and Cardiovascular Surgery.* 2010 Summer;22(2): pp. 127–39.

Chakir, K. and D. A. Kass (2010). Rethinking resynch: exploring mechanisms of cardiac resynchronization beyond wall motion control. *Drug Discovery Today: Disease Mechanisms* 7(2): pp. e103–7.

Chakravarti, A. and A. Kapoor (2012). Mendelian puzzles. *Science* 335(6071): pp. 930–1.

Chen, F., H. Kook, et al. (2002). Hop is an unusual homeobox gene that modulates cardiac development. *Cell* 110(6): pp. 713–23.

Chen, K. and N. Rajewsky (2006). Natural selection on human microRNA binding sites inferred from SNP data. *Nature Reviews Genetics* 38(12): pp. 1452–6.

Chen, Y., S. Park, et al. (2003). Alterations of gene expression in failing myocardium following left ventricular assist device support. *Physiological Genomics* 14(3): pp. 251–60.

Chen, Y., J. Zhu, et al. (2008). Variations in DNA elucidate molecular networks that cause disease. *Nature* 452(7186): pp. 429–35.

Chien, K. R. (2000). Myocyte survival pathways and cardiomyopathy: implications for trastuzumab cardiotoxicity. *Seminars in Oncology* 27(6 Suppl 11): pp. 9–14.

Chien, K. R. (2006). Herceptin and the heart—a molecular modifier of cardiac failure. *New England Journal of Medicine* 354(8): pp. 789–90.

Chinnery, P. F. *Mitochondrial Disorders Overview.*

Chua, T. P. and A. J. S. Coats (1995). The lungs in chronic heart failure. *European Heart Journal* 16(7): pp. 882–7.

Cleland, J., N. Freemantle, et al. (2008). Predicting the long-term effects of cardiac resynchronization therapy on mortality from baseline variables and the early response: a report from the

CARE-HF (Cardiac Resynchronization in Heart Failure) trial. *Journal of the American College of Cardiology* 52(6): pp. 438–45.

Coats, A. J. S. (2001). What causes the symptoms of heart failure? *Heart* 86(5): pp. 574–8.

Cohn, J. N. (1995). Structural basis for heart failure: ventricular remodeling and its pharmacological inhibition. *Circulation* 91(10): pp. 2504–7.

Colbert, M. (2002). Retinoids and cardiovascular developmental defects. *Cardiovascular Toxicology* 2(1): pp. 25–39.

Cowie, M. R., A. Mosterd, et al. (1997). The epidemiology of heart failure. *European Heart Journal* 18(2): pp. 208–25.

Creemers, E. E., A. A. Wilde, et al. (2011). Heart failure: advances through genomics. *Nature Reviews Genetics* 12(5): pp. 357–62.

Czech, M. P. (2006). MicroRNAs as therapeutic targets. *New England Journal of Medicine* 354(11): pp. 1194–5.

Danziger, R. S., M. You, et al. (2005). Discovering the genetics of complex disorders through integration of genomic mapping and transcriptional profiling. *Current Hypertension Reviews* 1(1): pp. 21–34.

Darbar, D. (2010). Genomics, heart failure and sudden cardiac death. *Heart Failure Reviews* 15(3): pp. 229–38.

Deng, M. C., H. J. Eisen, et al. (2006). Noninvasive discrimination of rejection in cardiac allograft recipients using gene expression profiling. *American Journal of Transplantation* 6(1): pp. 150–60.

Desruennes, M. l., T. Corcos, et al. (1988). Doppler echocardiography for the diagnosis of acute cardiac allograft rejection. *Journal of the American College of Cardiology* 12(1): pp. 63–70.

Dhitavat, J., L. Dode, et al. (2003). Mutations in the sarcoplasmic/endoplasmic reticulum $Ca2+$ ATPase isoform cause Darier's disease. *Journal of Investigative Dermatology* 121(3): pp. 486–9.

Donahue, M. P., D. A. Marchuk, et al. (2006). Redefining heart failure: the utility of genomics. *Journal of the American College of Cardiology* 48(7): pp. 1289–98.

Donnelly, P. (2008). Progress and challenges in genome-wide association studies in humans. *Nature* 456(7223): pp. 728–31.

Drexler, H., U. Riede, et al. (1992). Alterations of skeletal muscle in chronic heart failure. *Circulation* 85(5): pp. 1751–9.

Dunselman, P. H. J. M., C. E. E. Kuntze, et al. (1988). Value of New York Heart Association classification, radionuclide ventriculography, and cardiopulmonary exercise tests for selection of patients for congestive heart failure studies. *American Heart Journal* 116(6 Part 1): pp. 1475–82.

Eberhard, M. and P. Erne (1994). Calcium and magnesium binding to rat parvalbumin. *European Journal of Biochemistry* 222(1): pp. 21–6.

Eichelbaum, M., M. Ingelman-Sundberg, et al. (2006). Pharmacogenomics and individualized drug therapy. *Annual Review of Medicine* 57(1): pp. 119–37.

Esquela-Kerscher, A. and F. J. Slack (2006). Oncomirs—microRNAs with a role in cancer. *Nature Reviews Cancer* 6(4): pp. 259–69.

Esteller, M. (2008). Epigenetics in cancer. *New England Journal of Medicine* 358(11): pp. 1148–59.

Esther, C. R., Jr., S. Hoberman, et al. (2011). Detection of rapidly growing mycobacteria in routine cultures of samples from patients with cystic fibrosis. *Journal of Clinical Microbiology* 49(4): pp. 1421–5.

Ewer, M. S. and S. M. Lippman (2005). Type II chemotherapy-related cardiac dysfunction: time to recognize a new entity. *Journal of Clinical Oncology* 23(13): pp. 2900–2.

Fatkin, D., B. K. McConnell, et al. (2000). An abnormal Ca2+ response in mutant sarcomere protein-mediated familial hypertrophic cardiomyopathy. *Journal of Clinical Investigation* 106(11): pp. 1351–9.

Fehlbaum, P., C. Guihal, et al. (2005). A microarray configuration to quantify expression levels and relative abundance of splice variants. *Nucleic Acids Research* 33(5): p. e47.

Felker, G. M., K. F. Adams Jr, et al. (2003). The problem of decompensated heart failure: nomenclature, classification, and risk stratification. *American Heart Journal* 145(2 Suppl): pp. S18–25.

Finck, B. N., J. J. Lehman, et al. (2002). The cardiac phenotype induced by PPARα overexpression mimics that caused by diabetes mellitus. *Journal of Clinical Investigation* 109(1): pp. 121–30.

Force, T., D. S. Krause, et al. (2007). Molecular mechanisms of cardiotoxicity of tyrosine kinase inhibition. *Nature Reviews Cancer* 7(5): pp. 332–44.

Francis, G. S., K. M. McDonald, et al. (1993). Neurohumoral activation in preclinical heart failure. Remodeling and the potential for intervention. *Circulation* 87(5 Suppl): pp. IV90–6.

Freemantle, N., P. Tharmanathan, et al. (2006). Cardiac resynchronisation for patients with heart failure due to left ventricular systolic dysfunction—a systematic review and meta-analysis. *European Journal of Heart Failure* 8(4): pp. 433–40.

Frey, N., T. Barrientos, et al. (2004). Mice lacking calsarcin-1 are sensitized to calcineurin signaling and show accelerated cardiomyopathy in response to pathological biomechanical stress. *Nature Medicine* 10(12): pp. 1336–43.

Gardiner, S. J. and E. J. Begg (2005). Pharmacogenetic testing for drug metabolizing enzymes: is it happening in practice? *Pharmacogenetics and Genomics* 15(5): pp. 365–9.

Giralder, A. J., R. M. Cinalli, et al. (2005). MicroRNAs regulate brain morphogenesis in zebrafish. *Science* 308: pp. 833–8.

Goldstein, D. B. (2009). Common genetic variation and human traits. *New England Journal of Medicine* 360(17): pp. 1696–8.

Gomes, A., I. Falcão-Pires, et al. Rodent models of heart failure: an updated review. *Heart Failure Reviews*: pp. 1–31.

Grenier, M. A. and S. E. Lipshultz (1998). Epidemiology of anthracycline cardiotoxicity in children and adults. *Seminars in Oncology* 25(4 Suppl 10): pp. 72–85.

Guttmacher, A. E. and F. S. Collins (2002). Genomic medicine—a primer. *New England Journal of Medicine* 347(19): pp. 1512–20.

Hall, J. L., S. Grindle, et al. (2004). Genomic profiling of the human heart before and after mechanical support with a ventricular assist device reveals alterations in vascular signaling networks. *Physiological Genomics* 17(3): pp. 283–91.

Hare, J. M., G. D. Walford, et al. (1992). Ischemic cardiomyopathy: endomyocardial biopsy and ventriculographic evaluation of patients with congestive failure, dilated cardiomyopathy and coronary artery disease. *Journal of the American College of Cardiology* 20(6): pp. 1318–25.

Heidecker, B., E. K. Kasper, et al. (2008). Transcriptomic biomarkers for individual risk assessment in new-onset heart failure. *Circulation* 118(3): pp. 238–46.

Horwitz, P. A., E. J. Tsai, et al. (2004). Detection of cardiac allograft rejection and response to immunosuppressive therapy with peripheral blood gene expression. *Circulation* 110(25): pp. 3815–21.

Hosenpud, J. D. (1992). Noninvasive diagnosis of cardiac allograft rejection. Another of many searches for the grail. *Circulation* 85(1): pp. 368–71.

Hoshijima, M. and K. R. Chien (2002). Mixed signals in heart failure: cancer rules. *Journal of Clinical Investigation* 109(7): pp. 849–55.

Hovnanian, A. (2008). Serca Pumps and Human Diseases. *Calcium Signalling and Disease*. E. Carafoli and M. Brini, Springer Netherlands. 45: pp. 337–63.

Hu, Z., B. Yang, et al. (2008). HSPB2/MKBP, a novel and unique member of the small heat-shock protein family. *Journal of Neuroscience Research* 86(10): pp. 2125–33.

Hughes, S. E. and W. J. McKenna (2005). New insights into the pathology of inherited cardiomyopathy. *Heart* 91(2): pp. 257–64.

Hunt, S. A., D. W. Baker, et al. (2001). ACC/AHA guidelines for the evaluation and management of chronic heart failure in the adult: executive summary: A report of the American College of Cardiology/American Heart Association Task Force on Practice Guidelines (Committee to Revise the 1995 Guidelines for the Evaluation and Management of Heart Failure) developed in collaboration with the International Society for Heart and Lung Transplantation endorsed by the Heart Failure Society of America. *Journal of the American College of Cardiology* 38(7): pp. 2101–13.

Hunter, J. J. and K. R. Chien (1999). Signaling pathways for cardiac hypertrophy and failure. *New England Journal of Medicine* 341(17): pp. 1276–83.

Ino-Oka, E., Y. Kutsuwa, et al. (2001). Evaluation of the severity of chronic heart failure by the reactivity of peripheral vessels. *The Tohoku Journal of Experimental Medicine* 195(1): pp. 1–10.

Jefferies, J. L. and J. A. Towbin (2010). Dilated cardiomyopathy. *The Lancet* 375(9716): pp. 752–62.

Jessup, M. and S. Brozena (2003). Heart failure. *New England Journal of Medicine*. 2003 May 15;348(20): pp. 2007–18.

Joensuu, H., P.-L. Kellokumpu-Lehtinen, et al. (2006). Adjuvant docetaxel or vinorelbine with or without trastuzumab for breast cancer. *New England Journal of Medicine* 354(8): pp. 809–20.

Jooss, K. and N. Chirmule (2003). Immunity to adenovirus and adeno-associated viral vectors: implications for gene therapy. *Gene Therapy* 10(11): pp. 955–63.

Jung, K. and R. Reszka (2001). Mitochondria as subcellular targets for clinically useful anthracyclines. *Advanced Drug Delivery Reviews* 49(1–2): pp. 87–105.

Kakadekar, A. P., G. G. S. Sandor, et al. (1997). Differences in dose scheduling as a factor in the etiology of anthracycline-induced cardiotoxicity in Ewing sarcoma patients. *Medical and Pediatric Oncology* 28(1): pp. 22–6.

Kasowski, M., F. Grubert, et al. Variation in transcription factor binding among humans. *Science* 328(5975): pp. 232–5.

Katz, A. M. (2008). The "modern" view of heart failure. *Circulation: Heart Failure* 1(1): pp. 63–71.

Kaya, Z., C. Leib, et al. (2012). Autoantibodies in heart failure and cardiac dysfunction. *Circulation Research* 110(1): pp. 145–58.

Keulenaer, G. W. and D. L. Brutsaert (2007). Pathophysiology and Clinical Impact of Diastolic Heart Failure. *Cardiovascular Medicine*. J. T. Willerson, H. J. J. Wellens, J. N. Cohn and D. R. Holmes, Springer London: pp. 1201–15.

Khoury, M. J. (2003). Genetics and genomics in practice: The continuum from genetic disease to genetic information in health and disease. *Genetics in Medicine* 5(4): pp. 261–8.

Kim, E. K. and E.-J. Choi (2010). Pathological roles of MAPK signaling pathways in human diseases. *Biochimica et Biophysica Acta (BBA)—Molecular Basis of Disease* 1802(4): pp. 396–405.

Kittleson, M. M. and J. M. Hare (2005). Molecular signature analysis: using the myocardial transcriptome as a biomarker in cardiovascular disease. *Trends in Cardiovascular Medicine* 15(4): pp. 130–8.

Kittleson, M. M., K. M. Minhas, et al. (2005). Gene expression analysis of ischemic and nonischemic cardiomyopathy: shared and distinct genes in the development of heart failure. *Physiological Genomics* 21(3): pp. 299–307.

Kittleson, M. M., S. Q. Ye, et al. (2004). Identification of a gene expression profile that differentiates between ischemic and nonischemic cardiomyopathy. *Circulation* 110(22): pp. 3444–51.

Kumar, D. (2011). The personalised medicine: a paradigm of evidence-based medicine. *Annali dell'Istituto Superiore di Sanità* 47: pp. 31–40.

Kwon, C., Z. Han, et al. (2005). MicroRNA1 influences cardiac differentiation in *Drosophila* and regulates Notch signaling. *Proceedings of the National Academy of Sciences of the United States of America* 102(52): pp. 18986–91.

Levine, T., G. Francis, et al. (1983). The neurohumoral and hemodynamic response to orthostatic tilt in patients with congestive heart failure. *Circulation* 67(5): pp. 1070–5.

Li, Y. Y., C. F. McTiernan, et al. (2000). Interplay of matrix metalloproteinases, tissue inhibitors of metalloproteinases and their regulators in cardiac matrix remodeling. *Cardiovascular Research* 46(2): pp. 214–24.

Liew, C.-C. and V. J. Dzau (2004). Molecular genetics and genomics of heart failure. *Nature Reviews Genetics* 5(11): pp. 811–25.

Liew, C. C. (2004). Heart Failure: A Genomics Approach. *Proteomic and Genomic Analysis of Cardiovascular Disease*, Wiley-VCH Verlag GmbH & Co. KGaA: pp. 45–60.

Liggett, S. B. (2010). Pharmacogenomics of beta1-adrenergic receptor polymorphisms in heart failure. *Heart Failure Clinics* 6(1): pp. 27–33.

Limongelli, G., M. Tome-Esteban, et al. (2010). Prevalence and natural history of heart disease in adults with primary mitochondrial respiratory chain disease. *European Journal of Heart Failure* 12(2): pp. 114–21.

Lin, K. C., K. Gyenai, et al. (2006). Candidate gene expression analysis of toxin-induced dilated cardiomyopathy in the Turkey (*Meleagris gallopavo*). *Poultry Science* 85(12): pp. 2216–21.

Lipshultz, S. E. (2006). Exposure to anthracyclines during childhood causes cardiac injury. *Seminars in Oncology* 33: pp. 8–14.

Liu, P. P. and J. W. Mason (2001). Advances in the understanding of myocarditis. *Circulation* 104(9): pp. 1076–82.

Maass, A. H., K. Ikeda, et al. (2004). Hypertrophy, fibrosis, and sudden cardiac death in response to pathological stimuli in mice with mutations in cardiac troponin T. *Circulation* 110(15): pp. 2102–9.

Machado, R. D., O. Eickelberg, et al. (2009). Genetics and genomics of pulmonary arterial hypertension. *Journal of the American College of Cardiology* 54(1 Suppl S): pp. S32–42.

MacRae, C. (2010). The genetics of congestive heart failure. *Heart Failure Clinics* 6(2): pp. 223–30.

Mann, D. L., A. Deswal, et al. (2002). New therapeutics for chronic heart failure. *Annual Review of Medicine* 53(1): pp. 59–74.

Manning, G., D. B. Whyte, et al. (2002). The protein kinase complement of the human genome. *Science* 298(5600): pp. 1912–34.

Manohar, P. and I. L. Piña (2003). Therapeutic role of angiotensin II receptor blockers in the treatment of heart failure. *Mayo Clinic Proceedings* 78(3): pp. 334–8.

Maradia, K. and M. Guglin (2009). Pharmacologic prevention of anthracycline-induced cardiomyopathy. *Cardiology in Review* 17(5): 243–52 10.1097/CRD.0b013e3181b8e4c8.

Margulies, K. B., D. P. Bednarik, et al. (2009). Genomics, transcriptional profiling, and heart failure. *Journal of the American College of Cardiology* 53(19): pp. 1752–9.

Maron, B. J., J. A. Towbin, et al. (2006). Contemporary definitions and classification of the cardiomyopathies: an American Heart Association Scientific Statement from the Council on Clinical Cardiology, Heart Failure and Transplantation Committee; Quality of Care and Outcomes Research and Functional Genomics and Translational Biology Interdisciplinary Working Groups; and Council on Epidemiology and Prevention. *Circulation.* 2006 Apr 11;113(14): pp. 1807–16. Epub 2006 Mar 27.

Mayosi, B. M., A. Kardos, et al. (2006). Heterozygous disruption of SERCA2a is not associated with impairment of cardiac performance in humans: implications for SERCA2a as a therapeutic target in heart failure. *Heart* 92(1): pp. 105–9.

McCarthy, M. I., G. R. Abecasis, et al. (2008). Genome-wide association studies for complex traits: consensus, uncertainty and challenges. *Nature Reviews Genetics* 9(5): pp. 356–69.

McCarty, D. M. (2008). Self-complementary AAV vectors; advances and applications. *Molecular Therapy* 16(10): pp. 1648–56.

McKinsey, T. A. and E. N. Olson (2004). Cardiac histone acetylation—therapeutic opportunities abound. *Trends in Genetics* 20(4): pp. 206–13.

McLeod, H. L. and W. E. Evans (2001). Pharmacogenomics: unlocking the human genome for better drug therapy. *Annual Review of Pharmacology and Toxicology* 41(1): pp. 101–21.

Meijering, R. A., D. Zhang, et al. (2012). Loss of proteostatic control as a substrate for atrial fibrillation: a novel target for upstream therapy by heat shock proteins. *Front Physiol* 3(36): p. 23.

Michaels, P. J., M. L. Espejo, et al. (2003). Humoral rejection in cardiac transplantation: risk factors, hemodynamic consequences and relationship to transplant coronary artery disease. *The Journal of Heart and Lung Transplantation: The Official Publication of the International Society for Heart Transplantation* 22(1): pp. 58–69.

Minotti, G., E. Salvatorelli, et al. (2010). Pharmacological foundations of cardio-oncology. *Journal of Pharmacology and Experimental Therapeutics* 334(1): pp. 2–8.

Molkentin, J. D., J.-R. Lu, et al. (1998). A calcineurin-dependent transcriptional pathway for cardiac hypertrophy. *Cell* 93(2): pp. 215–28.

Monsuez, J.-J., J.-C. Charniot, et al. (2010). Cardiac side-effects of cancer chemotherapy. *International Journal of Cardiology* 144(1): pp. 3–15.

Morita, H., J. Seidman, et al. (2005). Genetic causes of human heart failure. *Journal of Clinical Investigation* 115(3): pp. 518–26.

Most, P., A. Remppis, et al. (2007). S100A1: a novel inotropic regulator of cardiac performance.

Transition from molecular physiology to pathophysiological relevance. *American Journal of Physiology—Regulatory, Integrative and Comparative Physiology* 293(2): pp. R568–77.

Na Bangchang, K., W. Songsaeng, et al. (1994). Pharmacokinetics of primaquine in G6PD deficient and G6PD normal patients with vivax malaria. *Transactions of the Royal Society of Tropical Medicine and Hygiene* 88(2): pp. 220–2.

Noutsias, M., S. Pankuweit, et al. (2009). Biomarkers in inflammatory and noninflammatory cardiomyopathy. *Herz* 34(8): pp. 614–23.

O'Donnell, J. M. and E. D. Lewandowski (2005). Efficient, cardiac-specific adenoviral gene transfer in rat heart by isolated retrograde perfusion in vivo. *Gene Therapy* 12(12): pp. 958–64.

Olson, R. and P. Mushlin (1990). Doxorubicin cardiotoxicity: analysis of prevailing hypotheses. *The FASEB Journal* 4(13): pp. 3076–86.

Omori, K. and J. Kotera (2007). Overview of PDEs and their regulation. *Circulation Research* 100(3): pp. 309–27.

Ordovas, J. M. and C. E. Smith (2010). Epigenetics and cardiovascular disease. *Nature Reviews Cardiology* 7(9): pp. 510–9.

Osterziel, K. J. and A. Perrot (2005). Dilated cardiomyopathy: more genes means more phenotypes. *European Heart Journal* 26(8): pp. 751–4.

Pattison, J., J. Waggoner, et al. (2008). Phospholamban overexpression in transgenic rabbits. *Transgenic Research* 17(2): pp. 157–70.

Periasamy, M. and P. M. L. Janssen (2008). Molecular basis of diastolic dysfunction. *Heart Failure Clinics* 4(1): pp. 13–21.

Pham, M. X., J. J. Teuteberg, et al. (2010). Gene-expression profiling for rejection surveillance after cardiac transplantation. *New England Journal of Medicine* 362(20): pp. 1890–900.

Piccart-Gebhart, M. J., M. Procter, et al. (2005). Trastuzumab after adjuvant chemotherapy in HER2-positive breast cancer. *New England Journal of Medicine* 353(16): pp. 1659–72.

Plana, J. C. (2011). Chemotherapy and the heart. *Revista Española de Cardiología* 64(05): pp. 409–15.

Porębska, I., A. Harłozińska, et al. (2000). Expression of the tyrosine kinase activity growth factor receptors (EGFR, ERB B2, ERB B3) in colorectal adenocarcinomas and adenomas. *Tumor Biology* 21(2): pp. 105–15.

Qiu, H., H. Dai, et al. (2008). Characterization of a novel cardiac isoform of the cell cycle-related kinase that is regulated during heart failure. *Journal of Biological Chemistry* 283(32): pp. 22157–65.

Quaini, F., K. Urbanek, et al. (2002). Chimerism of the transplanted heart. *New England Journal of Medicine* 346(1): pp. 5–15.

Ramay, H. R., O. Z. Liu, et al. (2011). Recovery of cardiac calcium release is controlled by sarcoplasmic reticulum refilling and ryanodine receptor sensitivity. *Cardiovascular Research* 91(4): pp. 598–605.

Robertson, K. D. (2005). DNA methylation and human disease. *Nat Rev Genet* 6(8): pp. 597–610.

Rodenhiser, D. (2009). Epigenetic contributions to cancer metastasis. *Clinical and Experimental Metastasis* 26(1): pp. 5–18.

Rose, B. A., T. Force, et al. (2010). Mitogen-activated protein kinase signaling in the heart: angels versus demons in a heart-breaking tale. *Physiological Reviews* 90(4): pp. 1507–46.

Rose, J., A. A. Armoundas, et al. (2005). Molecular correlates of altered expression of potassium currents in failing rabbit myocardium. *American Journal of Physiology: Heart and Circulatory Physiology* 288(5): p. 6.

Sandor, G. G. S., M. Puterman, et al. (1992). Early prediction of anthracycline cardiomyopathy using standard M-mode and digitized echocardiography. *Journal of Pediatric Hematology/Oncology* 14(2): pp. 151–7.

Sayed, D., S. Rane, et al. (2008). MicroRNA-21 targets sprouty2 and promotes cellular outgrowths. *Molecular Biology of the Cell* 19(8): pp. 3272–82.

Schoels, M., T. J. Dengler, et al. (2004). Detection of cardiac allograft rejection by real-time PCR analysis of circulating mononuclear cells. *Clinical Transplantation* 18(5): pp. 513–7.

Schott, J.-J., D. W. Benson, et al. (1998). Congenital heart disease caused by mutations in the transcription factor NKX2-5. *Science* 281(5373): pp. 108–11.

Schramm, C., D. M. Fine, et al. (2012). The PTPN11 loss-of-function mutation Q510E-Shp2 causes hypertrophic cardiomyopathy by dysregulating mTOR signaling. *American Journal of Physiology: Heart and Circulatory Physiology* 302(1): p. 4.

Schrier, R. W. and W. T. Abraham (1999). Hormones and hemodynamics in heart failure. *New England Journal of Medicine* 341(8): pp. 577–85.

Schuelke, M., K. R. Wagner, et al. (2004). Myostatin mutation associated with gross muscle hypertrophy in a child. *New England Journal of Medicine* 350(26): pp. 2682–8.

Segers, V. F. M. and R. T. Lee (2008). Stem-cell therapy for cardiac disease. *Nature* 451(7181): pp. 937–42.

Seidman, J. G. and C. Seidman (2001). The genetic basis for cardiomyopathy: from mutation identification to mechanistic paradigms. *Cell* 104(4): pp. 557–67.

Seo, D. and P. Goldschmidt-Clermont (2008). Cardiovascular genetic medicine: the genetics of coronary heart disease. *Journal of Cardiovascular Translational Research* 1(2): pp. 166–70.

Shah, R. R. (2004). Pharmacogenetic aspects of drug-induced torsade de pointes: potential tool for improving clinical drug development and prescribing. *Drug Safety* 27(3): pp. 145–72.

Shull, G. E., G. Okunade, et al. (2003). Physiological functions of plasma membrane and intracellular Ca2+ pumps revealed by analysis of null mutants. *Annals of the New York Academy of Sciences* 986(1): pp. 453–60.

Sile, S., C. G. Vanoye, et al. (2006). Molecular physiology of renal ClC chloride channels/

transporters. *Current Opinion in Nephrology and Hypertension* 15(5): pp. 511–6 10.1097/01.mnh.0000242177.36953.be.

Singhal, S., D. Miller, et al. (2008). Gene expression profiling of non-small cell lung cancer. *Lung Cancer* 60(3): pp. 313–24.

Smith, I., M. Procter, et al. (2007). 2-year follow-up of trastuzumab after adjuvant chemotherapy in HER2-positive breast cancer: a randomised controlled trial. *The Lancet* 369(9555): pp. 29–36.

Sparano, J. A. (1998). Use of dexrazoxane and other strategies to prevent cardiomyopathy associated with doxorubicin-taxane combinations. *Seminars in Oncology* 25(4 Suppl 10): pp. 66–71.

Steenbergen, C., C. A. Afshari, et al. (2003). Alterations in apoptotic signaling in human idiopathic cardiomyopathic hearts in failure. *American Journal of Physiology—Heart and Circulatory Physiology* 284(1): pp. H268–76.

Sternick, E. B., A. Oliva, et al. (2011). Clinical, electrocardiographic, and electrophysiologic characteristics of patients with a fasciculoventricular pathway: the role of PRKAG2 mutation. *Heart Rhythm* 8(1): pp. 58–64.

Takemura, G. and H. Fujiwara (2007). Doxorubicin-induced cardiomyopathy: from the cardiotoxic mechanisms to management. *Progress in Cardiovascular Diseases* 49(5): pp. 330–52.

Tan, C. D., W. M. Baldwin, et al. (2007). Update on cardiac transplantation pathology. *Archives of Pathology & Laboratory Medicine* 131(8): pp. 1169–91.

Tanaka, M., C. I. Berul, et al. (2002). A mouse model of congenital heart disease: cardiac arrhythmias and atrial septal defect caused by haploinsufficiency of the cardiac transcription factor Csx/Nkx2.5. *Cold Spring Harbor Symposia on Quantitative Biology* 67: pp. 317–26.

Tatsuguchi, M., H. Y. Seok, et al. (2007). Expression of microRNAs is dynamically regulated during cardiomyocyte hypertrophy. *Journal of Molecular and Cellular Cardiology* 42(6): pp. 1137–41.

Taylor, D. O., M. L. Barr, et al. (1999). A randomized, multicenter comparison of tacrolimus and cyclosporine immunosuppressive regimens in cardiac transplantation: decreased hyperlipidemia and hypertension with tacrolimus. *The Journal of Heart and Lung Transplantation* 18(4): pp. 336–45.

Thum, T., C. Gross, et al. (2008). MicroRNA-21 contributes to myocardial disease by stimulating MAP kinase signalling in fibroblasts. *Nature* 456(7224): pp. 980–4.

Toda, N. (2003). Vasodilating β-adrenoceptor blockers as cardiovascular therapeutics. *Pharmacology & Therapeutics* 100(3): pp. 215–34.

Toko, H., W. Zhu, et al. (2002). Csx/Nkx2-5 is required for homeostasis and survival of cardiac myocytes in the adult heart. *Journal of Biological Chemistry* 277(27): pp. 24735–43.

Tonks, N. K. (2006). Protein tyrosine phosphatases: from genes, to function, to disease. *Nature Reviews Molecular Cell Biology* 7(11): pp. 833–46.

Towbin, J. and N. Bowles (2000). Genetic abnormalities responsible for dilated cardiomyopathy. *Current Cardiology Reports* 2(5): pp. 475–80.

Trivedi, C. M., Y. Luo, et al. (2007). Hdac2 regulates the cardiac hypertrophic response by modulating Gsk3[beta] activity. *Nature Medicine* 13(3): pp. 324–31.

Valantine, H. (2004). Cardiac allograft vasculopathy after heart transplantation: risk factors and management. *The Journal of Heart and Lung Transplantation: The Official Publication of the International Society for Heart Transplantation* 23(5): pp. S187–93.

van Rooij, E., L. B. Sutherland, et al. (2006). A signature pattern of stress-responsive microRNAs that can evoke cardiac hypertrophy and heart failure. *Proceedings of National Academy of Sciences* 103(48): pp. 18255–60.

van Rooij, E., L. B. Sutherland, et al. (2007). Control of stress-dependent cardiac growth and gene expression by a MicroRNA. *Science* 316(5824): pp. 575–9.

van Vliet, J., N. Oates, et al. (2007). Epigenetic mechanisms in the context of complex diseases. *Cellular and Molecular Life Sciences* 64(12): pp. 1531–8.

Velagaleti R. S., O. D. C. (2010). Genomics of heart failure. *Heart Failure Clinics* 6(1): pp. 115–24.

Vinge, L. E., P. W. Raake, et al. (2008). Gene therapy in heart failure. *Circulation Research.* 2008 Jun 20;102(12): pp. 1458–70.

Vogel, C. L., M. A. Cobleigh, et al. (2002). Efficacy and safety of trastuzumab as a single agent in first-line treatment of HER2-overexpressing metastatic breast cancer. *Journal of Clinical Oncology* 20(3): pp. 719–26.

Walter, S. D., L. Irwig, et al. (1999). Meta-analysis of diagnostic tests with imperfect reference standards. *Journal of Clinical Epidemiology* 52(10): pp. 943–51.

Wang, X. and X. Wang (2006). Systematic identification of microRNA functions by combining target prediction and expression profiling. *Nucleic Acids Research* 34(5): pp. 1646–52.

Wehrens, X. H. T., S. E. Lehnart, et al. (2005). Intracellular calcium release and cardiac disease. *Annual Review of Physiology* 67(1): pp. 69–98.

Whitney, A. R., M. Diehn, et al. (2003). Individuality and variation in gene expression patterns in human blood. *Proceedings of National Academy of Sciences* 100(4): pp. 1896–901.

Wienholds, E., W. P. Kloosterman, et al. (2005). MicroRNA expression in zebrafish embryonic development. *Science* 309(5732): pp. 310–1.

Wilk, J. B., R. H. Myers, et al. (2006). Adrenergic receptor polymorphisms associated with resting heart rate: the HyperGEN study. *Annals of Human Genetics* 70(5): pp. 566–73.

Williams, M. L. and W. J. Koch (2004). Viral-based myocardial gene therapy approaches to alter cardiac function. *Annual Review of Physiology* 66(1): pp. 49–75.

Williams, P., P. Ranjzad, et al. (2010). Development of viral vectors for use in cardiovascular gene therapy. *Viruses* 2(2): pp. 334–71.

Wittstein, I. S., D. R. Thiemann, et al. (2005). Neurohumoral features of myocardial stunning due to sudden emotional stress. *New England Journal of Medicine* 352(6): pp. 539–48.

Yasuda, S. and W. Y. Lew (1997). Lipopolysaccharide depresses cardiac contractility and beta-adrenergic contractile response by decreasing myofilament response to Ca2+ in cardiac myocytes. *Circulation Research* 81(6): pp. 1011–20.

Zhang, C. L., T. A. McKinsey, et al. (2002). Class II histone deacetylases act as signal-responsive repressors of cardiac hypertrophy. *Cell* 110(4): pp. 479–88.

Zhao, Y., E. Samal, et al. (2005). Serum response factor regulates a muscle-specific microRNA that targets Hand2 during cardiogenesis. *Nature* 436(7048): pp. 214–20.

Zhao, Y. and D. Srivastava (2007). A developmental view of microRNA function. *Trends in Biochemical Sciences* 32(4): pp. 189–97.

Zsáry, A., S. Szûcs, et al. (2004). Endothelins: a possible mechanism of cytostatics-induced cardiomyopathy. *Leukemia & Lymphoma* 45(2): pp. 351–5.